T0073098

THE SOCIAL ASPECTS OF ENVIRONMENTAL AND CLIMATE CHANGE

The Social Aspects of Environmental and Climate Change critically examines the prominence of natural science framing in mainstream climate change research and demonstrates why climate change really is a social issue.

The book highlights how assumptions regarding social and cultural systems that are common in sustainability science have impeded progress in understanding environmental and climate change. The author explains how social sciences theory and perspectives provide an understanding of institutional dynamics including issues of scale, possibilities for learning, and stakeholder interaction, using specific case studies to illustrate this impact. The book highlights the foundational role research into social, political, cultural, behavioural, and economic processes must play if we are to design successful strategies, instruments, and management actions to act on climate change.

With pedagogical features such as suggestions for further reading, text boxes, and study questions in each chapter, this book will be an essential resource for students and scholars in sustainability, environmental studies, climate change, and related fields.

E. Carina H. Keskitalo is Professor of Political Science at the Department of Geography, Umeå University, Sweden.

Routledge Advances in Climate Change Research

Climate Change Law in China in Global Context
Edited by Xiangbai He, Hao Zhang, and Alexander Zahar

The Ethos of the Climate Event
Ethical Transformations and Political Subjectivities
Kellan Anfinson

Perceptions of Climate Change from North India
An Ethnographic Account
Aase J. Kvanneid

Climate Change in the Global Workplace
Labour, Adaptation and Resistance
Edited by Nithya Natarajan and Laurie Parsons

Governing Climate Change in Southeast Asia
Critical Perspectives
Edited by Jens Marquardt, Laurence L. Delina and Mattijs Smits

Kick-Starting Government Action against Climate Change
Effective Political Strategies
Ian Budge

The Social Aspects of Environmental and Climate Change
Institutional Dynamics Beyond a Linear Model
E. C. H. Keskitalo

For more information about this series, please visit: www.routledge.com/Routledge-Advances-in-Climate-Change-Research/book-series/RACCR

THE SOCIAL ASPECTS OF ENVIRONMENTAL AND CLIMATE CHANGE

Institutional Dynamics Beyond a Linear Model

E. C. H. Keskitalo

Cover credit: Orbon Alija @ iStock

First published 2022
by Routledge
2 Park Square, Milton Park, Abingdon, Oxon OX14 4RN

and by Routledge
605 Third Avenue, New York, NY 10158

Routledge is an imprint of the Taylor & Francis Group, an informa business

British Library Cataloguing-in-Publication Data
A catalogue record for this book is available from the British Library

Library of Congress Cataloging-in-Publication Data
Names: Keskitalo, E. C. H. (Eva Carina Helena), 1974– author.
Title: The social aspects of environmental and climate change : institutional dynamics beyond a linear model / E.C.H. Keskitalo.
Description: Abingdon, Oxon ; New York, NY : Routledge, 2022.
Series: Routledge advances in climate change research | Includes bibliographical references and index.
Identifiers: LCCN 2021039479 (print) | LCCN 2021039480 (ebook) | ISBN 9780367489977 (hardback) | ISBN 9780367489960 (paperback) | ISBN 9781003043867 (ebook)
Subjects: LCSH: Climatic changes—Social aspects. | Environmental sociology.
Classification: LCC QC903 .K45537 2022 (print) | LCC QC903 (ebook) | DDC 304.2/5—dc23
LC record available at https://lccn.loc.gov/2021039479
LC ebook record available at https://lccn.loc.gov/2021039480

ISBN: 978-0-367-48997-7 (hbk)
ISBN: 978-0-367-48996-0 (pbk)
ISBN: 978-1-003-04386-7 (ebk)

DOI: 10.4324/9781003043867

Typeset in Bembo
by Apex CoVantage, LLC

CONTENTS

FIGURES

TEXT BOXES

ACKNOWLEDGEMENTS

This book, for my part, is to a great extent a continuation and explication of the social environmental research work I have been doing in the past decades, such as in a handbook on climate change adaptation policy research, work on long-term change in resource sectors, and others.

The brevity of the nature of this book – intended for students as well as scholars in highly varying fields – means that it takes up only a part of the manifold, and far broader, work related to these issues. Thanks are due to all those who are struggling to secure recognition of social science and humanities research relevant to environmental change – among funders, decision-makers, and other scientists who have not yet become used to relating to these highly complex and varying fields.

For taking the time to read and offer suggestions on the manuscript, thanks are also due to the editors and anonymous reviewers, as well as to my department colleague Cenk Demiroglu for his encouragement.

1

AIM AND SCOPE OF THE BOOK

Introduction

How a research problem is framed is of crucial importance.

In the policy research field, there is a great deal of research asserting that framing a problem is actually the first – and often most crucial – step in determining how to solve it, what actors are important for solving it, and how this should be done (e.g. Schön and Rein 1994).

As a result, many of the struggles in the policy arena involve framing. For instance, should a flood event be seen as a result of climate change or just as a freak accident? If it is framed as a result of climate change, this might mean that there is a need for more far-reaching actions, for instance flood-proofing policy, avoiding building in low-lying areas, and the like. If it is framed as a freak accident, however, the focus might be placed more on direct emergency response after the event, with no need for costly additional actions or inclusion in policy. These types of storylines can thereby serve to promote different understandings of a problem (Hajer 1995).

Thus,

> to frame is to *select some aspects of a perceived reality and make them more salient . . . in such a way as to promote a particular problem definition, causal interpretation, moral evaluation, and/or treatment recommendation* for the item described.
>
> *(Entman 1993: 52, quoted in Weaver 2007: 143, italics in original; see also Reese 2007; Zhou and Moy 2007)*

In the same way, the argument in this work is that the framing in research plays a large role. If an issue is seen mainly as one of gaining better knowledge about the environment per se and improved technology, the focus will be placed on this:

DOI: 10.4324/9781003043867-1

better natural science knowledge, stronger technology development. However, if it is also seen as one of how present systems can manage implementing the decisions needed (for instance, for adapting to climate change or limiting invasive species spread), the focus will be another.

Some of these issues of how framing occurs may result in long-term and potentially costly misframings. For instance, the focus on gaining better knowledge about the environment per se has been strongly emphasised in the framing in the largest scientific assessment on climate change, the Intergovernmental Panel on Climate Change (IPCC) reports. Largely, the assumption has been that knowledge will lead to action. However, in reports from both 2007 and 2014 (see e.g. IPCC 2007, 2014) it is finally acknowledged that "simply producing more and better knowledge is not sufficient" (Mimura et al. 2014: 887) for achieving policy change.

This understanding, and a criticism of the notion that better scientific knowledge will automatically lead to better policies, is a crucial focus in this book. The assumption that better scientific knowledge will automatically lead to better policies is a crucial component in the disproven linear model of expertise, or linear model of scientific knowledge as it will be referred to here.[1] This is a complex of assumptions in which mainly natural science knowledge and technological development are seen as automatically leading to change, in a way that is rational from a scientific viewpoint (see Text Box 1.1). It has been recognised as regularly applied in large-scale research assessment programmes and assessments, as quite simply "the way science is practised".

TEXT BOX 1.1 THE LINEAR MODEL OF SCIENTIFIC KNOWLEDGE

"Linear models of scientific knowledge transfer have been labelled as 'technocratic'; they propose that scientific knowledge can be directly applied to practice without any problems or changes. Science helps to increase the effectiveness and efficiency of practical decisions, and scientific knowledge leads to better policies and 'evidence-based' policy making. If scientific facts are available, then this availability automatically leads to the use of science in practice. In linear models, political actors demand scientific knowledge, and it is supplied by the science community: this knowledge flows directly from science into practice, in which it is utilised by political decision makers and stakeholders with the aim of producing 'rational' solutions. Decisions are then based on scientific information without being altered by the policy process or practical considerations. The linear model of scientific knowledge transfer is still highly relevant among natural or climate scientists who, critically, ask why political actors or society as a whole do not follow their recommendations" (Böcher and Krott 2014: 3643).

Critics note that the pervasive application of the assumptions in the linear model of scientific knowledge needs to be explicitly identified and problematised, as it otherwise leads to acting on incorrect assumptions regarding how society works. Criticism of the linear model has focused on the fact that it ignores the notion that science cannot usefully be conceived of without also understanding society (Beck and Mahony 2018; Briggle 2008).[2] Critics also highlight that scientific knowledge, in actual life, is not automatically taken up and translated into action; this is because, among other things, policy-makers have to be responsive to very different types of input of which science is only a single, and often relatively minor, actor (De Koning et al. 2014). Real-world analyses have shown that scientifically "rational" decisions are not the most rational ones for actors who may instead follow their self-interests; this means that the scientific results may not be seen as useful or meaningful in relation to their situation (Böcher and Krott 2014). Rather than a notion that gaining use of knowledge may only be about removing "barriers" to scientifically "rational" decisions, policy processes and the actors within them may thus have entirely different goals, understandings of pertinent issues, and motivations (Grundmann 2009).

The linear model thus rests on an incorrect understanding of the policy process. The fact that the model – despite this criticism – is still used has led different authors to see it as "undead" (Durant 2015: 17), as an "ideology" (Grundmann 2009: 402), or as a "myth" (Briggle 2008; Pielke 2007).

There have also been attempts to rework the model. To attempt to address the concern that science does not take into account social considerations, a focus on including stakeholder participation has often been added to assumptions that are in line with the linear model in scientific programmes and assessments (Beck 2011; Durant 2015; Pielke 2007). However, also this addition carries with it numerous considerations. Just like the linear model assumes a removal from society, participation that simply assumes direct and clear deliberation with stakeholders – that they would in fact act similarly to how politicians are assumed to act in the linear model, working with science in a clear and "rational way" focused on scientific priorities – has proven to be incorrect (e.g. Philips 1995). Criticism of this type of focus includes, among other things, the assertion that real-life deliberation is never unconstrained and that those arguing for it in a simplified, unproblematised way may instead increase the influence of already powerful actors, without reflecting many actual societal divisions and consequences of decisions (Durant 2015; Keskitalo and Preston 2019).

Despite all this criticism, this model of a mainly natural science development that today includes societal aspects by means of stakeholder participation, for the purposes of decision support, remains in almost general application (Beck 2011; Durant 2015).

> *This book takes its point of departure in the notion that this misunderstanding of the social, political, and economic world in this way remains a major – and perhaps the largest – obstacle to correctly understanding social environmental research and its contributions.*

This is because, in fact, in the faulty linear model, there is no need to understand social, economic, and political considerations – institutional dynamics – through research, as the social, economic, and political actors are assumed to automatically relate to and implement best (natural science) knowledge. But because they do not do this, the model does not work and the processes that apply this model are left without the means to integrate the research on social, economic, and political actors and worlds that are in fact needed to conceive of change.

In this, the climate change field is by no means an exception. Climate change adaptation and mitigation, as well as environmental problems in general, are crucially issues of framing. The field of climate change research has largely been formed by its origins in impacts-based scenario research, and it is still often undertaken in the format of, for instance, integrated assessments or vulnerability assessments, which often focus on natural science components with added stakeholder participation, largely using modelling and scenario-building as decision-support tools (e.g. De la Vega-Leinert et al. 2008; De la Vega-Leinert and Schroter 2009; Noble et al. 2014; Beck and Mahony 2018; Malone and Engle 2011 for an overview). At the same time, there has been a criticised consistently low involvement of the social sciences and the humanities (Victor 2015; Stern and Dietz 2015) – and as noted earlier, only recently has the IPCC, based on this data, started clarifying that knowledge is not enough.[3]

What this book does

This book challenges the assumption that only a limited range of social sciences and humanities research would be relevant for understanding environmental issues such as adaptation and mitigation to climate change – including why mitigation and adaptation often do not take place.

Climate change mitigation regularly refers to what has to be done to limit emissions, while *adaptation* refers to what has to be done to deal with the changes that will take place – even if we were able to halt emissions today.[4] Adaptation thus includes, for instance, strategies and actions for dealing with changes in temperature and precipitation such as seasonal shifts, changes in how much rain we get when and where, and extreme events like storms and floods. It also includes recognizing and responding to the effects of these sorts of changes, such as changes in production (e.g. food production), impacts on water, pest outbreaks, or the spread of invasive species (IPCC 2014).

Thus, adaptation will be needed almost everywhere: such as in how to build to withstand floods, dimension sewer and water pipes, provide electricity even under storms, and produce food in a changing climate. Therefore, while the focus in adaptation research has often been on the local level, the role of adaptation is by no means limited to a local focus. Instead, these types of actions need to be decided upon and undertaken in virtually all governance systems, making adaptation an intensely social and complex question (e.g. Keskitalo and Preston 2019).

While many examples in the book will, for these reasons, be taken from the adaptation field, the discussions here regarding the role of the social system and transitions within it are equally applicable to mitigation as well as to broader environmental problems. While examples will be taken from, for instance, assessment work that summarises the literature, the treatment of social issues in these examples can be seen as symptomatic of larger fields such as "sustainability science", energy research, integrated assessment, and the like (e.g. Sovacool 2014). The book's focus is thereby not as much on understanding only one issue, or criticising only one literature or report, as it is on the fact that framings that build on the linear model are pervasive to multiple bodies of literature. Literature on climate change is an example, as are specific reports, but the linear model can be found in multiple bodies of literature with the same problems. The hope is that, through the illustration of these issues, the reader will become aware of the components of the linear model and be able to distinguish these in whatever field they occur in – and will then also be able to contrast them with other, more social framings.

To this end, *the book itself thus constitutes a "framing", directly contradicting the framing in the still-influential linear model of scientific knowledge.* The framing in this book focuses on understanding the real-life institutions that need to respond to, and have indeed created, the climate change problem. The focus is placed on understanding climate change adaptation (and, by implication, mitigation and broader environmental issues) through a framing focused on institutions and institutional aspects.

The book can thereby be seen as providing an institutionalist framing of the problem of adaptation, with the assumption that if this logic can be understood, climate change and environmental issues will also be seen and analysed in a different way (cf. Beck 2011). The aim is to forward this framing and what difference it makes in viewing climate change adaptation compared to how it is often done in literature (e.g. Victor 2015; Stern and Dietz 2015; Noble 2019).

On this basis, the volume illustrates that adaptation and mitigation considerations are necessarily implemented in an existing social structure with historically established institutions, and that it is through understanding the breadth of research on these topics that we can understand how to design strategies, instruments, and management actions for the future.

Understanding climate change and society – or environment and society – is thereby not only about limits and barriers, or other means that to some extent black-box societal processes (e.g. IPCC 2007, 2014). Rather, it is about understanding the foundational role played by research on social, political, cultural, behavioural, economic, and political processes in understanding the institutions through which adaptation and mitigation need to take place – if they are to happen at all.

Understanding this means that literature and approaches relevant to understanding – for instance, mitigation and adaptation in a social context – include not only research on climate change per se. They also include research on the institutions, including the processes and incentive structures through which adaptation and mitigation actions would need to be designed and implemented (cf. Young 1999).

In addition, the book asserts that this type of understanding of the huge contribution the social sciences can make to climate change or environmental issues also involves understanding the nature of the social sciences and the humanities. In these research fields, answers to questions about the institutional structure for responding to change are often described in theoretical perspectives: These include knowledge from multiple case studies abstracted into principles or process analyses, for instance regarding what is needed for a development to take place (e.g. Ruddin 2006). Understanding this broad range of research and theoretical perspectives, as well as the existing but also delimited transferability of knowledge, is crucial to understanding climate change in a social context.

The framing in this book thereby contrasts the assumptions in a linear model of scientific knowledge and associated assumptions, with broad framings that in some way focus on institutional aspects, while providing examples of what this could entail. In line with the broad focus on understanding framings, institutions and institutional aspects are also understood in a broad way here. Generally, social theory is relatively united in the notion that individuals do not act in a vacuum and cannot be seen apart from larger social, economic, and political contexts (King 2004).[5] What are seen here as institutions are highlighted to focus on this aspect of *real-life conditions* and the larger socioeconomic and political contexts that form the context for individual action. The focus in this also includes *higher scales* than the individual or local bottom-up perspective that has often been in focus in climate change adaptation (to be discussed in Chapter 2). This context may then be referred to in many ways, and while it is referred to here as institutional, the basic gist of the argument could be agreed on by many theoretical orientations but with great variation in how it is described, what terms are used to describe it, and what levels are in focus.[6]

For these reasons, rather than aiming for a comprehensive review of various institutionalisms – as a social scientist might understand the focus on institutions – the book's focus is instead placed on shifting the framing towards what understanding climate change adaptation at all (and by implication environmental issues) through real-life institutions, in real-life situations, could entail. This book does not aim to be a textbook in institutionalism, in that it will not cover all the different theories of institutionalism; nor does it aim to limit itself to theories of institutionalism. Instead, the focus is to discuss the system that needs to respond to climate change and environmental issues as one composed of institutions (cf. Thornton et al. 2012). To this end, examples to demonstrate this and the different logic it implies will be taken from a broad range of theories that nevertheless relate to what could be called "institutions" – but the aim is not to be comprehensive with regard to covering all types of relevant theories.[7]

This book thus:

- takes its basis in the notion that environmental problems are social problems and focuses on the social facts and real-life considerations necessary to understand environmental problems;

- proceeds from the position that a linear scientific knowledge transfer model is wrong, not only because it makes incorrect assumptions about policy advice but also because it exempts the social sciences and social science research problems as research fields and thereby misdescribes social, economic, political, and related issues, including questions regarding knowledge transfer, participation, and stakeholder integration; and
- proceeds from the assumption that what society can do to, for instance, adapt and mitigate is itself a research problem and should be treated as such, with use of the research tools available (and applied) to do so in the social sciences (see further Text Box 1.2).

TEXT BOX 1.2 WHAT THIS BOOK DOES NOT DO, IN RELATION TO WHAT IT DOES DO

- This book only takes a position on social science in, for instance, broader assessment work; it does not aim to involve with natural science facts per se, but only with how social facts are treated. This means that the book does not include a focus on the co-production of scientific claims or the ways in which, for instance, IPCC knowledge per se is construed. This itself is a large body of literature, beyond the specific factors placed in focus in relation to social science. The text thus takes no issue whatsoever with the notion that natural science must understand the physical and environmental processes of how much we need to mitigate/adapt, such as the degree goals and natural disruptions that take place if we don't meet them.

  Rather, the position here is that when we get to *how* to adapt and mitigate, this is largely an issue of understanding social, political, and economic systems and how they function.

  In line with a focus on the idea that the best science is always needed, the focus in this book is that we should go to the social sciences to answer that *how* question, as social systems constitute the research problems in the social sciences.
- The book is not explicitly concerned with developing any one set, formal model for science-policy interaction per se (in relation to the available and plentiful literature in this area). The book is thereby not about a "science-policy integration model" per se – rather, it is about how to get the best (and, implicitly, social – as adaptation and mitigation are largely social) science for supporting an understanding of *how* to adapt to and mitigate climate change.

  However, the position that social science is crucial to understand society at all naturally has implications on how to design science-policy interaction (as this has been crucial for motivating the addition of stakeholder participation to the linear model of scientific knowledge).

- The book does not (at all) aim to exclude a focus in relevant areas of the humanities from the focus on social sciences – rather, a crucial focus in the book is on the great importance of particularly historical studies in understanding structures, institutions, and possibilities for change. While the book does not develop an argument clearly across the full scope of the humanities, its focus is on the importance of including all sciences and research relevant to understanding the broadly conceived social problem at hand.

Compared with what the implications of implementing a linear model of scientific knowledge or expertise have been, the implications of the framing of environmental problems expressed in this work are significant:

- Taking the position that everything that goes beyond impact is to a very great extent social means that social science research is crucial to dealing with environmental problems – perhaps even more crucial today than the natural sciences, in the climate change case. This is because we already know relatively well the environmental problem we face, but to a much lesser degree have we collected and in a coherent way drawn conclusions from even existing knowledge concerning how to manage climate change as related to adaptation and mitigation.
- Taking the position here that social systems are not irrational, chaotic, wicked problems or un-understandable, but instead extremely complex (with this reflected in the research on these systems), may mean a total reversal in how decision support is undertaken. It may mean starting from assessing the motivations and needs among "stakeholders" through social study – rather than assuming a need – and then tailoring research to the instruments that can support this, rather the other way around (that is, not starting with developing scientific knowledge that is assumed to be taken up). It also means that it is not enough to highlight complexity and uncertainty while black-boxing social decision-making (e.g. Olsson et al. 2015; Wellstead et al. 2013). Rather, the position here is that the best existing science, i.e. that for which the social world is the main research focus, should be used to understand how to develop and support action on environmental change.
- Science-based decision support could look much different than it does today. Providing decision support related to potential future changes has so far largely been the purvey of developing scenarios based on modelling. It is well recognised that these seldom constitute "projections" – as it is difficult to project the future – but rather "most probable developments" or quite simply illustrations to support decision-makers (e.g. Von Storch et al. 2011). With the same

caveat – probable or less probable developments – understandings of social systems' historical and present patterns, particularly institutionalised actor groups/organisations, could be used to provide relevant decision support. While there are caveats to any assumption that the future could in any way be foretold (path dependency-type perspectives as similar to any understanding of scenarios as predictions), this does not mean that the output of computer-based models constitutes any more likely "most probable developments" than do broader institutional studies. There is thus no basis for using only the former, as is regularly done.

Outline

The book will firstly define the problem and pervasiveness of the linear model of scientific knowledge with examples from the climate change field and critique the social understanding expressed in this literature (Chapter 2).

Secondly, the book will discuss the role of theory in the social sciences and the broad range of institutional perspectives that can provide an understanding of why we are not sufficiently adapting to or mitigating climate change. The chapter will discuss the potential to effect change through, for instance, policy instruments and the possibility to provide "solutions" to what are essentially political problems of effecting change (Chapter 3).

As an application of what taking a broad institutional perspective could mean, Chapter 4 introduces examples of the application of such a perspective in cases of forestry and climate change adaptation, such as multilevel adaptation governance. The chapter provides a basis for a return to these examples throughout the book, in subsequent chapters, with applications on the issues treated there.

The three following chapters highlight specific assumptions in relation to expressions of the linear model of scientific knowledge in literature and why they are incorrect or limited. Chapter 5 discusses why knowledge and learning mechanisms for including knowledge are not enough to conceive of social context for a "knowledge transmission". Chapter 6 discusses how power influences any approach to participation or stakeholder interaction (and why it is not enough to simply speak with stakeholders, as well as why people can be "rational" in different ways). Chapter 7 then illustrates how a focus on smaller-scale traditional communities, which has often been prevalent in the literature, may obscure some other understandings of societies and explains why taking the same actions and achieving the same results at different localities may not be possible. A final chapter on the implications for understanding possibilities for change as well as conducting research concludes the book.

Key points

- The linear model of scientific knowledge sees science as apart from society, and it assumes that science will be able to develop knowledge that results in societal change.

- Additions to this model to manage criticism, such as a role for stakeholder participation and the need for decision support for knowledge use, have not fundamentally changed the basic assumptions in the linear model to include understandings of society in any more integrated way.
- Instead, seeing environmental problems as social problems – with the need to integrate social perspectives with an institutional component in understandings of stakeholders, systems, and framings of climate change – has major implications for how scientific practice and assessment are undertaken.

Study questions

- What is the linear model of scientific knowledge?
- What has the criticism of the linear model of scientific knowledge entailed?
- Why do you think the linear model of scientific knowledge is still used, even though it has been disproven?
- What are the consequences of using the linear model for framing environmental issues?

Notes

1 The term "linear model of scientific knowledge" is used here to refer to what is sometimes also called the linear model of expertise, which has been identified to share characteristics with the linear model of innovation (e.g. Durant 2015). To indicate the potentially very large areas of thought that this type of model thereby refers to, and to highlight the knowledge transfer implied, the term "the linear model of scientific knowledge" (or just "the linear model") is generally used to refer to this model in this book.

2 This model thus attempts to separate "science" and "society" from each other: something that has been purported as an attempt at being value-free (Beck and Mahony 2018), but has been seen as resulting in an "excess of objectivity" as it does not correctly recognise that different disciplines only highlight their specific choices of issues and thereby cannot reflect the complex world in its entirety (Briggle 2008). The attempt to separate "science" from "society" is perhaps also related to the fact that the "science" placed in focus is that of a generalist natural science, for which all-encompassing natural laws are sought, and for which the variation in human societies, culture, politics, and economics has not been placed in focus (cf. Snow 1961; Henderson 2003; Hollis and Smith 1991). Omitting the policy process as well as social processes – including, by implication, largely the research on them – the model and its applications simply do not match the real world.

3 The international scientific advancement regarding decision-making context in the IPCC has been limited by the recognised, and criticised, consistently low involvement of the social sciences and the humanities (Victor 2015; Stern and Dietz 2015). Basic knowledge in social sciences such as the role of knowledge in decision-making could have been made available to the IPCC as early as in 1990 for the first general assessment report (e.g. based in environmental psychology such as Proshansky et al. 1970; Tanner 1980; McAndrew 1993). Instead, the focus in natural science literature on improving knowledge as the key social mechanism (e.g. Noble et al. 2014) led to a more individual rather than structural focus, such as on social science studies emphasising stakeholders or social learning (Klein et al. 2014; Mimura et al. 2014; Noble et al. 2014; Collins and Ison 2009; cf. Brulle and Dunlap 2015).

4 In the 2014 IPCC AR5 report, adaptation is defined as "[t]he process of adjustment to actual or expected climate and its effects. In human systems, adaptation seeks to moderate

harm or exploit beneficial opportunities. In natural systems, human intervention may facilitate adjustment to expected climate and its effects" (Noble et al. 2014: 838). It is also noted that the term "seeks to moderate" refers to purposeful action.

5 As Manicas expresses it, "[u]nlike the objects of study in natural science, the objects of study in social science – institutions, social structures, social relations – do not exist independently of us. They are, as we shall explain, real but concept- and activity-dependent" (Manicas 2006: 43). That is, how we understand our surrounding world influences and even constitutes it. If we act as if there is a world market, specific traditions to adhere to, or state structures, we are supporting their continued existence.

6 There are multiple, more detailed, descriptions and means of analysis, none of which mean that a general focus on what here is called institutions is wrong when it comes to real-life empirical analysis intended to inform decision-making; but many are complementary, more detailed, and use other terms. This is the nature of social study, as there are very many features one can choose to highlight and have then developed specific means to do so. It does not mean that even those focusing on the individual level do not agree that the individual level is composed of and constructed (or however one wants to conceptualise this) by numerous features outside of it (King 2004).

7 The focus is also not on promoting change or taking action per se, but on promoting an understanding of the problems of viewing climate change – or even the environment – from interdisciplinary viewpoints.

Additional readings

Hulme, M. (2009) *Why We Disagree about Climate Change: Understanding Controversy, Inaction and Opportunity*. Cambridge University Press, Cambridge.

2

THE ORIGINS OF CLIMATE CHANGE RESEARCH AND ASSUMPTIONS WITHIN THE FIELD

Introduction

The previous chapter introduced the issues around the linear model of expertise or scientific knowledge as an underlying assumption in much climate change and environmental literature. This chapter will, firstly, summarise main concerns involving this model, including the fact that purely gaining better knowledge is not enough to gain policy action. The chapter also highlights the problem that it is not necessarily the case that this better knowledge can be communicated and applied directly by stakeholders. For these reasons, it is crucial that knowledge also include knowledge – i.e. research – about society.

Secondly, the chapter will discuss how this model has manifested in, for instance, IPCC assessment reports, but is by no means limited to this literature. The argument here is that it is important to be able to identify the parts of the linear model in action. To this end, the chapter will discuss the origins of climate change research in impacts research, among other areas, and the consequences this has had on framing and steering climate change research. The chapter thereby sets the stage for an identification and discussion of the limitations, assumptions, and gaps in present climate change research as well as more broadly natural science-based research on the environment, in which social science has to date played a very limited role (e.g. Victor 2015).

Conclusions of this chapter include that an important part of a social framing is to understand that actors will not necessarily prioritise an issue – such as climate change – just because it has the highest risk over the long term. If stakeholders or social actors do not have incentives that make it apparent why a certain issue should be prioritised over others, it is likely that it will not be prioritised. As all stakeholders will not be able to absorb, or even be interested in absorbing, knowledge about climate per se, what research can do is to instead absorb knowledge about

DOI: 10.4324/9781003043867-2

"stakeholders" and foremost the system that situates them and influences what they can do. All stakeholders are not alike; rather, there are considerable power differentials and ways in which they can or cannot manage concerns within the existing policy and regulative environment. They also act within existing policy and regulative environments, whereby different constitutional and legal structures in different countries provide different possibilities for change (e.g. Haider and Morford 2004). The linear model is thereby leading the development of actions on the environment, adaptation, and mitigation in the wrong way – ignoring the actual social knowledge that is needed to understand social actors like us all. The chapter shows that understanding this does not mean giving up on change in the system; rather, it means taking change – and understanding the actors among whom it has to be undertaken – seriously.

A summary of the problems with the linear model of scientific knowledge

- ***Better knowledge is not enough to gain policy action.***

As noted in the previous chapter, a foundational assumption in the linear model is on developing better knowledge as a basis for policy action (see Text Box 1.1, Chapter 1).

But while more and better knowledge is necessary as a basis for decision-making, it has routinely been acknowledged in social science research that *better knowledge is not enough to gain policy action*. For instance, as a part of basic studies in environmental psychology, it is acknowledged that the linkage between knowledge, attitudes and beliefs, and actions (putting any such knowledge into practice) is not a given (Moser and Dilling 2011). For instance, people, even if they believe something (such as climate change or invasive species risks) is a risk, may not take action on it and are even able to hold mutually exclusive attitudes at the same time (ibid.). Thus, the fact that knowledge is not enough would not have needed to be included in IPCC reports until only relatively recently (IPCC 2007, 2014) – instead, this fact was already well known in the social sciences far before climate change became a prominent issue.[1]

This means that one major problem with the linear model is that knowledge cannot be assumed to directly result in action. Not only is this because, as environmental psychology among others have shown, people can hold several incommensurate ideas or understandings without acting on any of them but rather acting habitually or in response to direct social requirements (family and the like) (e.g. Moser and Dilling 2011); it is also because decision-makers and stakeholders will not be free to act in whatever ways they wish. There is international, European Union (EU), and national legislation and regulation; there are multiple competing interests; there are institutionalised ways of being and acting – and all of these serve to encase existing and historical interests and ways of acting (e.g. Piattoni 2009). Without understanding the institutionalised structure of incentives and motivations, as well as institutionalised assumptions, knowledge transmission and the design of specific instruments to support change within this context will not be possible (e.g. Keskitalo and Preston 2019a) (see Text Box 2.1).

TEXT BOX 2.1 EXAMPLES OF HOW THE EXISTING SITUATION IMPACTS THE EXTENT TO WHICH STAKEHOLDERS CAN ACT: DIFFICULTIES IN REGULATING INVASIVE SPECIES

The risk of an increase in invasive species under climate change can be taken as one example of an issue that is difficult to regulate, given how the institutional framework on the international level is currently set up. The increase in invasive species risk is largely a result both of globalisation having resulted in transports taking place over much larger areas and much more frequently than before and of potential invasive species that gain access this way being able to establish themselves as climate change may make it possible for species to survive in areas where they previously might not have been able to (Pettersson et al. 2016).

The main influence on how trade between countries is regulated is the World Trade Organization's framework, which states that trade barriers can only be erected for cases like this if the invasive species and paths of spread are identified and the species is proven dangerous. This means that requirements for limiting trade are rather high: Often, it takes time to identify that there is a problem (for instance, dying trees), to find the culprit (the specific species), to then figure out where it came from, and to then – if this is not already known in the literature, as the species may not have had these effects in its original environment – prove it to be dangerous (ibid.).

As a result, even if scientists have argued that all non-essential trade in live plants should cease, arguing for the need for precaution and that this trade is non-essential, as was done through the Montesclaros Declaration (Montesclaros 2011; cf. Hantula et al. 2014), the limitations suggested are in fact not possible under the existing international system.

As major trade routes, the WTO system and EU regulation – just to mention a few considerations – are built on free trade, a major consideration in managing issues of invasive species thus involves understanding limitations in what can and cannot be changed under present considerations under the WTO, existing suggestions for reform of the WTO, and similar issues. In order to understand why progress on this issue is so difficult, one must thus understand this framework and how it may impact national, regional, and local actors, many of whom may not even be aware that consequences they identify are resultant from international frameworks – which were not even set up in relation to invasive species per se.

- ***It is not possible to directly communicate knowledge to stakeholders and automatically achieve action.***

A second and related problem with the linear model is that it assumes that it will be possible to *directly communicate* the better knowledge to "stakeholders" with the express purpose that they make use of it (e.g. Pielke 2007). The framing of the linear model assumes that knowledge will be directly relevant and usable by "society", without science having to involve itself in understanding motivations or include concerns about usability or interest (Beck 2011). What is sometimes called "wave one" thinking (the general linear model, e.g. Pielke 2007) is thereby added to by "wave two" thinking, including democratic aspects for managing critique of the model without changing its core (cf. Beck 2011).

A focus on societal actors as "stakeholders" (cf. Keskitalo 2004) that does not clearly aim at understanding social context can thereby be seen as enabling continued use of the linear model by slotting societal actors into a relationship with science, thereby managing the critique from more democratic viewpoints without abandoning the model. Members of society as "stakeholders" are thereby assumed to be directly able to relate to larger natural science framings and to directly communicate their considerations through, for instance, science-led stakeholder interaction (e.g. De la Vega-Leinert et al. 2008).

This assumption has many side effects. As direct communication is simply assumed in the linear model, by omission it excludes all science that deals with society, motivations, interest, or use (cf. Pielke 2007). This has meant that it has often been possible to focus "stakeholder" interaction, aimed at developing stakeholder support/decision support, rather directly on modelling and visualisation (e.g. Mimura et al. 2014; Antunes et al. 2006) without clearly understanding the social constraints on "stakeholder" actions. Thereby, it typically does not focus on assessing stakeholder needs in the context of, for instance, the institutional framework in which stakeholders act, and whether potential decision support systems can fit into this stakeholder context (as noted already by Kates 1985; cf. Holm et al. 2013; Hackmann et al. 2014). It also does not acknowledge that "stakeholders" are very seldom able to express the entire context of what limits what they can do, as to them this is a natural, everyday context (e.g. Haider and Morford 2004; Manicas 2006).

This means that it is not sufficient to directly ask "stakeholders" about issues or science without also having developed an understanding of the social, economic, and political situations and the institutions that govern them. Local or even national stakeholders, depending on their position, may not always be aware of, for instance, international frameworks that may limit their possibilities for action. "Stakeholders" may thus not have complete knowledge of all the impacts on their situation and may also have naturalised their situation and thus not be able to express all impacts on it in a direct interaction that is itself uninformed about potential influences (see Text Box 2.2).

TEXT BOX 2.2 EXAMPLES OF HOW THE EXISTING SITUATION IMPACTS THE EXTENT TO WHICH STAKEHOLDERS CAN ACT: THE NEED TO WORK WITHIN THE ESTABLISHED MOTIVATING FRAMEWORK, FOR INSTANCE FOR LOCAL COUNCILS/MUNICIPALITIES

Much adaptation work has argued that adaptation needs to be local in order to take into account the varying local contexts (such as the different problems for low-lying versus mountain areas, and also the different resources or more broadly adaptive capacities in different areas) (e.g. Nalau et al. 2015). However, given that the local level is within the state and all other governance frameworks are within established governance structures – structures of authority – the local level is in fact to a relatively large extent steered in relation to what can be done (ibid.). This does not mean that the local level cannot make decisions, but that decisions will take place within already established contexts.

To date, climate change adaptation has often been voluntary rather than a mandatory requirement, which itself may place it low on the list of priorities – below all other considerations that are required by, for instance, law or by the state, e.g. in instructions or other policy relevant in the country (Keskitalo 2010). For such reasons, much research shows that municipalities or local councils find it difficult to prioritise climate change adaptation even under perceived risk (e.g. Nalau et al. 2015). Climate change and, prior to this, other environmental issues have often come in as the "new issue on the block" and are forced to compete with all other, often very worthy, issues for resources and attention (e.g. Kingdon 1995).

So it is not necessarily that the local actors do not want to act or do not see the urgency, but rather that climate change is often practically set to compete with other worthy issues and different party priorities, while requirements for working on climate change compared to other issues have often been low. How can a local council or municipality prioritise climate change over social welfare and jobs (unless it is very clear what the impacts of climate change will be, and that they are immediate)?

In line with suggestions in established social science literature on how issues at large may gain priority in decision-making (e.g. Kingdon 1995), it has been shown that, for instance, large events and associated framings focusing on climate (for instance, floods that are depicted in media or by other prominent actors as related to climate change) and governmental frameworks whereby, for instance, the local level is funded for selecting to work on climate change, have been able to provide a rationale for working on climate that local government can defend to its constituency (e.g. Keskitalo et al. 2012).

- ***Knowledge needs to include knowledge about society.***

A final integral part of a linear model that has been criticised involves *which science or research is made relevant*. As seen earlier, the assumption regarding what science is has rested on natural science and perhaps technology. In the linear model, science is to exist apart from society (e.g. Pielke 2007); it is not to target society. In this, *science on society* is thereby excluded: Social science and, for instance, human science methods and knowledge were omitted, as if in some way "contaminated" by their focus on society, with all its varying values, political processes, and economic interests (and without understanding that research on these phenomena is needed in order to understand rather than ignore them, and that the phenomena are not the same as the research) (Beck 2011). It was only in the "wave two" thinking regarding the need to somehow include society (but without abandoning the linear model in which natural science is given a specific role) that the social sciences and humanities were included, but then largely as a component related to science communication, for instance in stakeholder assessments or to communicate natural science (see e.g. Beck 2011; Pielke 2007; Malone and Engle 2011). "Wave two" in the critique of the linear model thus highlights the need for stakeholder usability of science, but it was not necessarily interpreted by those already working in this paradigm as a need to shift the model.

A crucial problem here, and the reason why some authors have noted that the linear model is "undead" (alive despite having been disproven), is that regularly, the model's limitations are not even recognised; nor is the model itself. Instead, it is rather assumed as the basic model for scientific programme development (cf. Reed et al. 2016; Arts et al. 2017). The assumptions of this still "undead" model thus continue to carry significant consequences for what models and theories or knowledge are applied in research programmes and research on different levels (Durant 2015). Large-scale programmes or "sustainability science" in, for instance, the EU but also in nationally funded programmes, often target integrated assessment-style studies, including a large modelling component that attempts to describe change over time in ecosystems and for society with regard to potential scenarios (or imagined/potential storylines) of societal development (e.g. De la Vega-Leinert and Schroter 2009; Malone and Engle 2011; Reed et al. 2016). In this respect, stakeholders are often involved in order to discuss scenarios and what they would result in, and results are developed, for instance, for decision support, e.g. to make decision-makers in specific cases (municipalities, regions) think "outside the box" and start including aspects like climate change or other targeted considerations (e.g. Beck 2011; Reed et al. 2016). However, despite including a societal component in terms of "stakeholder integration", they often do not leave the linear model behind. Instead, by extending the linear model towards a linkage to society, they highlight stakeholder communication but not necessarily the understanding of stakeholder contexts, as these are not included in the model.

This means that as the linear model – even its extended version including stakeholder involvement – makes the social into an issue of communication, it *continuously omits that "the social" in fact is a research problem*. It also omits, as has been shown, that a relevant point of departure for research would be to understand the legislative, policy, and practice constraints on and context of stakeholder action, which requires research based in a social and present – not only future – context. What does social science research identify as the major limiting regulations, policies, important actors, and interest groups who determine what to do on an issue, and what type of regulative and policy environment could knowledge about climate change slot into? Even more, what type of laws, regulations, or policy could be set up: In practice, how could the existing situation be modified by the use of specific instruments or incentives, to include, for instance, climate change issues? Social scientists, who make a living describing these legislative, policy, and other contexts, constitute a crucial go-between in articulating limitations in the legislative, policy, economic, and other environments (see e.g. Manicas 2006; Haider and Morford 2004). Based on such an analysis, modelling and scenarios could then be defined, as a turnaround from the present standard setup of large assessments whereby modelling constitutes the basis for knowledge assumed to be communicated with stakeholders. Despite the potential feasibility of such a reversed design, however, it is almost never put into practice. Instead, the assumptions inherent in the linear model have meant that the social sciences and humanities regularly constitute a smaller proportion of researchers, often come in later, and have not served as driving forces in environmental research (e.g. Victor 2015; Palsson et al. 2013): They are mainly slotted into the space enabled by an extension of the linear model that includes communication with "stakeholders" but does not abandon the idea that natural science is central and in most ways removed from society (e.g. Beck and Mahony 2018; Briggle 2008).

The consequences of the application of a linear model have thereby also included entry points for social science having been few and far between, and largely centred on communication and stakeholder involvement. Attempts at including a broader focus have developed in social vulnerability literature (e.g. Adger 2006), for instance, but in this aspect seldom take into account a wider range of theory and knowledge (e.g. Wellstead et al. 2014). These types of criticism have been leveraged in multiple fields (e.g. Wellstead et al. 2013). In relation to mitigation and energy research, the journal *Energy Research and Social Science* was launched by a high-profile mitigation researcher when his papers were seen by editors of energy journals as not falling within the journals' disciplinary framings. The article explaining the launch of the journal states:

> Social science related disciplines, methods, concepts, and topics remain underutilized, and perhaps underappreciated, in contemporary energy studies research. . . . [C]ontent analysis of 4444 research articles involving 9549 authors and 90,079 references (from a smaller subsample) published in three leading energy journals from 1999 to 2013 . . . [shows that] only 19.6 percent of authors reported training in any social science discipline, and less than

> 0.3 percent of authors reported disciplinary affiliations in areas such as history, psychology, anthropology, and communication studies. Only 12.6 percent of articles utilized qualitative methods and less than 5 percent of citations were to social science and humanities journals.
>
> *(Sovacool 2014: 1)*[2]

Later, this and another researcher reviewed funding of mitigation research internationally, concluding that only 0.12% of it had been awarded to the social sciences (Overland and Sovacool 2020).

The way the linear model of scientific knowledge manifests – examples from IPCC reports

Given the pervasiveness of a linear model, it is not surprising that it is also pervasive in the IPCC reports, as one of the world's largest scientific assessments. However, as noted previously, the application of the linear model is by no means limited to this; it is regularly applied in large-scale sustainability science or integrated assessment programmes, as quite simply "the way one does science".

This type of criticism in no way says anything about the quality of the natural science research on climate change that is the focus in the IPCC reports (and for which the reports have been designed), which is per definition the best in the world. It also says nothing about the quality of any science applied under broader sustainability science or integrated assessment paradigms – to the extent that these can be removed from societal considerations. What this criticism does say a great deal about, however, is the basis on which social mechanisms are described, both in these reports and in the broader literature and multiple fields they draw upon. Not least, in the climate case, this is because social mechanisms and the social context as described earlier have regularly not been the target of climate-focused studies, and those studies that have been included have been the relatively few that have focused on climate per se (compared to all the research existing on social, economic, and political considerations that are relevant to implementing climate change adaptation and mitigation).

While the problem of the IPCC's omission of the social has been noted by many (e.g. Victor 2015; Stern and Dietz 2015), particularly Beck (2011) has provided a clear summary of how IPCC reporting draws on the assumptions in the linear model, taking adaptation as an example. The fact that some time has now passed since Beck's study was published and that new IPCC reports have been launched since is itself not of import: Understanding the type of criticism that is forwarded will be relevant as long as the reporting – or even work in larger fields – can be identified to somehow build on the assumptions in the linear model. In this context, as also noted in Chapter 1, the aim thus remains to exemplify the underlying logic or framing applied, whereby, even if developments are made over time, we need to remain alert to whether the linear model continues to be applied, or whether any processes are put in place to limit its use.

In line with the problems of the linear model described here, Beck notes that a key concern in the IPCC reports is that the focus is on causal explanation focused on the natural science knowledge that is to be communicated to society in order to effect change. Beck writes that:

> [t]he IPCC approach is based on classic "basic" research which seeks to understand abstract causal processes, thereby abstracting from local contexts and specific definitions. . . . In this approach, vulnerability [how we are vulnerable to climate change] is viewed as the *end point* of analysis, as the residual outcome of climate change impacts minus adaptation.
>
> *(Beck 2011: 299)*

Vulnerability, or how we may be affected by climate change, is thereby seen as more of a residual outcome than a start for analysis. Beck highlights that this "analysis of problems assumes specific causal processes which are taken to be linear and involve only a single variable. . . . The initial organization and three-tier structure of the IPCC is characterized by the sequence: science → impacts → response" (Beck 2011: 300).

In Beck's understanding, this linear approach to explanation rests on an analytical division between science and society. In it, science develops decision support, e.g. in global models and scenario-based approaches that are assumed to be able to gain responses at various levels. In other words, science is intended to highlight the impacts, which society is then intended to respond to. Science is thereby about "speaking truth to power" (Figure 2.1), operationalised through the notion that basic natural science-based research will result in changes in policy (Figure 2.2).

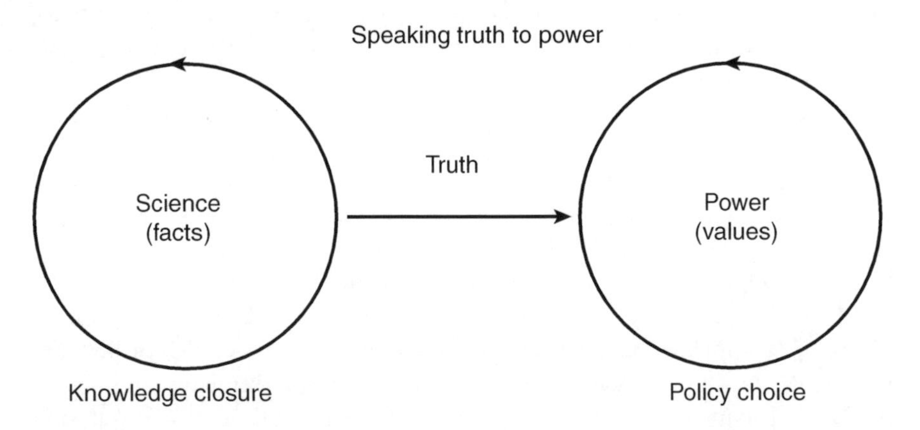

FIGURE 2.1 Speaking truth to power. Jasanoff and Wynne (1998: 9), reproduced from Beck 2011: 298.

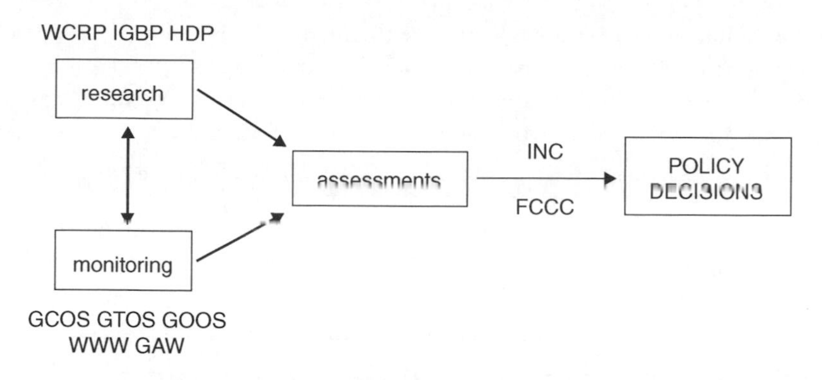

FIGURE 2.2 Science and policy-making. Bolin (1994: 27), reproduced from Beck 2011: 298.

Beck notes that this type of "use of the linear model . . . leads to . . . having an awareness of the political terrain while at the same time ignoring it" (Beck 2011: 299). It also leads to an incorrect understanding of adaptation as a far more limited issue than it is. Beck explains:

> The scenario-based approach was developed for the purpose of quantifying vulnerability to climate change, by asking what the extent of the climate change problem is and whether the costs of climate change exceed the costs of greenhouse gas mitigation. Following this approach, the IPCC focuses on specific predefined measurements of risk (such as the projected physical changes and impacts due to increasing greenhouse gas concentrations alone). Adaptation is thus framed as the degree to which adjustments are possible in practices, processes, and systemic structures in response to projected or actual climatic changes. Vulnerability summarizes the net impact of the climate problem and can be represented either quantitatively as monetary cost, human mortality, and ecosystem damage, or qualitatively as a description of relative or comparative change. Following this logic, the problem exists to the extent that mitigation policies fail. Adaptation is thus considered as a (marginal) cost of failed mitigation.
>
> *(Beck 2011: 300–301)*

The very real and actual political, social, and economic issues surrounding adaptation are thereby treated "scientifically" as more of an afterthought or a complexity: They become something that should take place as a result of the threat of climate change, with features then noted that may support or limit the risks of this. In total, this is what Beck means by saying that the political issue of climate change is depoliticised and abstracted, and that there is an awareness of the political while it is simultaneously ignored (Beck 2011). Social change, while central to the purpose of why science is undertaken, is assumed rather than researched, and the social is seen as something that is to change as a result of improved knowledge, rather than making the issue of how social systems change central to studies. The changes within politics are crucial, but they are assumed to take place as a response to improved knowledge, a development for which barriers could potentially be identified and removed.

Beck suggests that this type of framing of adaptation – not as an issue of how our governance or other systems work but as something that should take place if we are not able to mitigate sufficiently – has led to an implicit bias against adaptation policies. In this, it has also led to a marginalisation of adaptation and the politics of adaptation, resulting in a focus more on technical and top-down solutions than on the real costs, incentives, and justice and distributional issues for different groups (see Text Box 2.3).

TEXT BOX 2.3 BECK'S ANALYSIS OF HOW LINEAR MODEL ASSUMPTIONS LEAD TO A MISFRAMING OF ADAPTATION

"Proxy debates result in a *depoliticization of politics*. This occurs in the following ways.

First, as long as adaptation is discussed in terms of its marginal effects on anthropogenic climate change, its real importance for society is obscured and the political debate fails to address the core societal, political, and cultural problems involved. Restricting the focus to specific triggers of climatic change may not address the broader causes of vulnerability. Climatic changes and impacts are only part of the story. Resilience and vulnerability to climate-related impacts on society are increasing for reasons that have nothing to do with greenhouse gas emissions. As with [hurricane] Katrina, the political obsession with the idea that climate risks can be reduced by cutting emissions distracts attention from the more important factors that drive flood risks. As Hurricane Katrina made devastatingly clear, climate vulnerability is caused by unsustainable patterns of development combined with socio-economic inequity.

Second, following the linear model of expertise, politically relevant questions are framed and addressed in a very abstract, disembodied, and non-political way. Solutions often follow the well-trodden path of seeking a technological fix for a technologically created problem. The types of policy measures that emerge from these scenario-based assessments – such as irrigation schemes, drought tolerant seed varieties, and infrastructural improvements – are rather static and technical in nature.

Third, the linear model of expertise tends to stifle discussion of alternative policy approaches. Climate change is framed as a relatively 'tame' problem, which requires a straightforward solution, namely the top-down creation of a global carbon market. As a result, the broad spectrum of potential policy options is closed down in favor of a single option. . . . This also reinforces the marginalization of adaptation.

Fourth, . . . [there has been] reluctance to explain the concrete economic, social, and political implications of scientific findings. . . . As long as the IPCC remains reluctant to address the political implications of scientific findings, it does not meet the information needs of decision-makers. Decision-makers, however, need to know how climate change will affect specific political jurisdictions and, more importantly, what types of interventions will make a difference, over what time scales, at what costs, and to whose benefit. Many argue that this sterile approach may have made the IPCC less useful than it might otherwise have been".

(Beck 2011: 302–303)

Consequences of a linear framing of scientific knowledge: the use of concepts that limit the focus on decision-making

Beck's analysis shows that there are significant issues around assuming that vulnerability to climate change or adaptation to its effects can be assessed and managed without knowing the decision-making or institutional processes around it. As long as adaptation is assumed to take place more or less in relation to climate change knowledge, there is a limitation in understanding the extent to which social actors in fact act in relation to many other influences, how central (or not) climate change may be to them, and what factors influence social actors. This fact that "adaptation" needs to be understood in relation to much more than simply climate change is thereby under-pronounced.

This pervasiveness of the existing models for conceiving of what is relevant to climate change has also been noted by IPCC authors themselves. Noble et al., in the 2014 IPCC general assessment report, note that the " 'standard approach' to assessment has been the climate scenario-driven impacts-based approach" (Noble et al. 2014: 850): proceeding from a scenario-driven impacts-focused approach to response, as Beck (2011) criticises. In another publication, Noble later noted that the "IPCC was conceived of as a process for reporting recent advances in our understanding of climate change and its impacts largely in biophysical terms, rather than to explore, debate and develop a wider and more effective conceptual bases for progress" (Noble 2019: 43–44). Noble notes that this process was then supplemented with " 'second generation' vulnerability and adaptation assessments . . . characterized by the intensive involvement of stakeholders and the participation of vulnerable groups in decision-making around adaptation" (Noble et al. 2014: 851). However, it must also be noted that these second-generation assessments are akin to the "wave two" of criticism of the linear model: They respond to the argument that stakeholders are not sufficiently involved but do not per se shift the assumption regarding which sciences should be involved, or how to take into account the larger social and governance system and research on it in a coherent way.

In the same vein, the issue of how to conceive of adaptation – as a largely social, multilevel question – also remains. Noble notes that the 1990–2001 IPCC assessments included adaptation mainly as a "brief add-on" (Noble 2019: 30). While the inclusion of adaptation issues increased to a specific chapter in 2001, two in 2007, and four in 2014, he notes that "each was constrained within a negotiated, prescribed outline, which did not encourage a comprehensive interdisciplinary discussion of what is needed to achieve the actions necessary to progress adaptation" (Noble 2019: 30). While Noble et al. identify that there has been a shift between the fourth and fifth general assessment reports (2007–2014) of not only focusing on "responses to changes in biophysical systems" (Noble et al. 2014: 836), so far, any inclusion of a broader understanding of adaptation in social context is far from complete.

This framing based in the linear model could be seen as continuously resulting in a number of conceptions. These are treated in the following mainly through examples from the 2014 report which, as noted earlier, is the one that to date has most specifically treated adaptation as an issue.

- ***Social characteristics as largely removable discrete features***

In the 2014 IPCC report, the chapter "Adaptation opportunities, constraints, and limits", as reflected in the title, framed adaptation in relation to factors that could be seen as more favourable or more limiting. As a result, it thereby preferenced some conceptions of adaptation over others (Keskitalo and Preston 2019a). The chapter did not, for instance, take as its focus social science conceptions of the governance system, but instead it focused on the way varying features may limit or enable climate change, potentially describing "barriers" and "opportunities" (e.g., Adger et al. 2007; Klein et al. 2014). This type of conception was also echoed elsewhere in the report (e.g., Mimura et al. 2014).

The focus on barriers and opportunities has been criticised (e.g. Biesbroek et al. 2014). Much of the criticism notes that conceptualisations like "barriers" paint them more as removable obstacles than as the system features they may in fact be. Thus, while suggestions like "[C]oordination . . . can be a . . . barrier to planning and implementation" (Mimura et al. 2014: 881) are relevant, they underplay the fact that it may not be possible to remove "barriers" as they may be a part of how a system is designed.

Conceptualisations such as "barriers" or "opportunities" also do not necessarily mirror social literature; they thereby place the conceptualisations in the IPCC report aside from the context of the much larger bodies of research (beyond those focusing specifically on climate) that may already exist on the different types of features that could fall under these terms. In general, this focus has been seen to result in literature "remain[ing] mostly descriptive, and [classifying] rather than explain[ing] barriers" while "adaptation governance is much more than a technical procedure of adapting to 'given' changes: it is deeply and inherently political" (Oels 2019: 139).

In this way, "barriers" could be seen as a conception that relates to the linear model: If we can only remove barriers, it will be possible for knowledge to result in action. However, the reasons why something does not happen perhaps more seldom relate to barriers that can easily be removed, and perhaps more often to how a governance system is set up and what types of actions it currently incentivises. Text Boxes 2.1 and 2.2 in this chapter illustrate not only the limitations to what individual stakeholders can do in certain situations, but also that what could be called "barriers" are part of the systems and how they are organised – sometimes, "barriers" are related to the key features in an organisation. For instance, limitations in what can be done directly in regard to invasive species on an international level result from how the international system is currently set up. In relation to focal issues in these systems invasive species are only consequential, not central, making the trade limitations that do hinder direct action on invasive species difficult to change without larger change in potentially many other systems or areas (such as in a focus on free trade; see Text Box 2.1).

All this means that a focus must be placed on the social, political, and economic system in which "stakeholder considerations" and "barriers" are developed. This includes understanding the power of different actors and how the possibilities for different groups to act may already be embedded, and constrained or enabled in specific ways, in regulation and administration. This places the focus on what is possible to incrementally change as well as what different actors see as possible – not primarily in relation to any more general considerations on climate change but rather to laws and policy as well as the interests or constituents, such as voters, to whom they stand accountable (cf. Young 1999).

- ***Conceptions of social characteristics in blanket terms***

The use of terms like "barriers" and "opportunities" can be seen as an application of blanket terms to what may be arrays of different social features. Similarly, the focus in literature on understanding "vulnerability" and "adaptive capacity" can be seen as the application of highly general terms to cover a large array of social features that are already treated in different ways in existing social literature. Thus, it has been noted that "much of the research on adaptation policy opportunities, constraints, and limits has focused on access to capacities such as knowledge, capital, and resources" (Keskitalo and Preston 2019b: 8). This is a problem similar to the conception of "barriers": It makes features seem more easy to conceptualise, identify, manage, and eventually remove or enable than they in fact are. It can also be seen as conceptualising these types of features in relation to only some bodies of literature.

Beyond excluding much of the political process around adaptation and targeting only certain features, concepts such as "capacity" may result in larger assumptions of similarities between systems than what may actually be the case (see for instance Keskitalo et al. 2011; Keskitalo and Preston 2019b). While, for instance, economic parameters of adaptive capacity, or any other general parameters of this type, could be identified in most systems, the way they manifest in different cases may "differ so strongly that the term holds only very limited explanatory value" (Keskitalo and Preston 2019a: 490, footnote 3). In the social sciences, while different types of capacity exist as concepts, capacity conceptualisations are by far not the only means that are applied: They are often seen in the context of multiple other conceptualisations and as a delimited part of understanding, and they are not the only concepts used to conceive of possibilities to, for instance, respond in a situation (e.g. Wellstead et al. 2014).

Different from capacities, as features of a system, determining or even constituting a strong influence on actual action on climate change, authors have also noted that capacities cannot be regarded as sufficient in themselves but as something that has to be activated and mobilised, for instance within political processes (e.g. by Bay-Larsen and Hovelsrud 2017). Already relatively early in the adaptation field, literature clarified that existing institutional power structures have a strong influence on how responses are developed (Næss et al. 2005) and that it is also quite possible that existing structures are not well adapted in the present (e.g. Tompkins 2005). This means that taking into account climate change cannot be assumed even in cases in which broader "capacities" may exist or be possible to identify – in the same way that these cases may also not take into account other issues (see e.g. Text Box 2.2).

Thus, while adaptation has often been studied by applying the concept of "adaptive capacity", this conceptualisation may perhaps mostly reflect a hope that acting on climate change would be possible to understand and compare using single composite measurements, tools, or conceptualisations (cf. Wellstead et al. 2013, 2014), whereas in fact it relates to multiple and varying social systems.

- ***A focus on lower, individual or community, levels***

The reproduction in IPCC reports of considerations relevant to a linear model have also – taken together with the focus with which assessments were undertaken to target climate change specifically rather than broader social processes – reproduced a limited framing of social systems that largely excludes a focus on larger-scale social systems. This type of conception is by no means limited to IPCC reports, but it reproduces the assumptions in extensions to the linear model that target stakeholders and direct interaction. In relation to this, the social is largely conceptualised as individuals and stakeholders who act directly in relation to knowledge. While such stakeholders may act together, it is often assumed that they act on an individual level as part of smaller-scale communities where the direct interaction with scientific knowledge is to take place.

This type of conception has been seen as typical of the contrast between the large-scale, generalist, natural science aims and the requirement of "bottom-up" democratic interaction later introduced into the model. Thus, contrasting top-down and bottom-up tools for decision-making, planning, and implementation, the phrasing in one of the 2014 IPCC chapters is as follows: Top-down tools include downscaled simulated climate scenarios, while "[i]n the bottom-up approach, those affected or at risk examine their own impacts and vulnerabilities and incorporate adaptive options for the appropriate sector or community"

(Mimura et al. 2014: 883).[3] The extent to which broadly democratic modifications to the linear model are incorporated here relates to an inclusion of particularly traditional knowledge that is seen as relevant to smaller-scale communities. Thus, for instance, Mimura et al. note (in the IPCC general assessment report) that "local knowledge based adaptation is primarily focused on the use of traditional knowledge to increase adaptive capacity at the community level" (Mimura et al. 2014: 882) (see Text Box 2.4).

Conceptions like these thus largely assume that individual, community-level social actors make assessments based on climate change knowledge, translated as vulnerabilities and adaptive options. At the same time, issues higher than the local or community level are rather summarily treated, for instance in the IPCC 2014 general assessment report. While the role of institutions and governance is generally highlighted throughout the report, the accounts of these are more often descriptive or even prescriptive in suggesting what should be done (Keskitalo and Preston 2019a).[4] While larger-scale, "governance" features are mentioned, they are relatively sparsely described beyond abstractions, and the linkages between

TEXT BOX 2.4 COMMUNITY ADAPTATION – A HIGHLIGHTED BUT DELIMITED PART OF ADAPTATION

Community adaptation, as a prominent strain in adaptation literature, has largely focused on small-scale communities, where assessments can practically include all of the community and its variety of local interests (e.g. Ford et al. 2016; Hovelsrud and Smit 2010 eds). This focus can be seen as typical of the way a focus on the community level – which cannot be seen as standing apart from a focus on individual stakeholders developed in relation to the linear model – has been made prominent in climate change research.

However, there are notable problems with framing adaptation largely in the context of a community, without taking into account higher levels. Being able to undertake research only on a "community" scale – where it is possible to include all of the community in a detailed way – does not necessarily reflect larger considerations and higher levels, even in regions where such studies have been prominent (e.g. Keskitalo et al. 2011). Attempts to include a multi-level governance perspective in community adaptation research, for instance by reviewing the impacts of higher scale defined at the local level (e.g. Keskitalo 2008), indicate that it is not sufficient to view local (individual, group, and in some cases sector) decision-making (see also Text Box 2.2). Instead, local adaptation is already in some ways steered by higher-order decision-making as well as by how other priorities structure decision-making (O'Brien and Leichenko 2000) (see also Text Box 2.1).

"communities" and governance are not drawn out to any greater degree. Thereby, the framing does not clearly open up for discussing broad impacts on the decision-making process, such as potential power perspectives going into the development of adaptation options, limited resources to consider these, the power inherent in decision-making and the influences from higher organisational levels, or the potential competition between climate change and other issues (depending on how climate change issues are framed) (e.g. Wellstead et al. 2014).

By focusing on bottom-up interaction and on actors rather than systems – in the way systems may be understood in social science research – this type of conception excludes the major role social science knowledge could play, not in a stakeholder communication function but in providing knowledge regarding the various perspectives described earlier, which are not highlighted in a systematic way in the report. This results in a conception largely based on direct interaction with individuals or communities in an insufficiently described social context in which individuals are nevertheless assumed to learn from and come to apply scientific knowledge (Keskitalo and Preston 2019a).

- ***Social foci on learning and knowledge sharing***

In this connection, it may not be surprising that a number of statements in the IPCC 2014 general assessment report and other literature highlight the importance of learning as a knowledge transmission mechanism. For instance, it is noted that the "complexity of climate adaptation means that adaptation options are heavily influenced by forms of learning and knowledge sharing" (Noble et al. 2014: 847); that "[t]he increasing complexity of adaptation practice means that institutional learning is an important component of effective adaptation" (Noble et al. 2014: 836); that there is a "need for shared learning on adaptation" (ibid.: 842); and that "[a]daptation planning and implementation are dynamic iterative learning processes" (Mimura et al. 2014: 871).

One may note that the analytical assumptions between three concepts regularly placed in focus in adaptation literature in the report – *knowledge, learning*, and *capacity* – are "similar: they all take the existing system as given (rather than problematize it) and focus broadly on how it may transmit and learn knowledge and extend capacities" (Keskitalo and Preston 2019a: 490, footnote 3). The conception of stakeholder communication as a major role for the social sciences in the linear model can be traced here as well, resulting in the social sciences work that is made relevant being targeted towards lower-level (e.g. local level) stakeholder-focused work that places the role of knowledge transmission centrally (cf. Beck 2011).

Figure 2.3 attempts to illustrate this figuratively: If the focus is placed on knowledge transmission through learning, among largely unitary stakeholders (conceived without a clear conception of the broader social system), a problematisation of the "system" is largely left out. It thereby excludes inherently social science understandings of the systems, such as how stakeholders relate to each other, what the

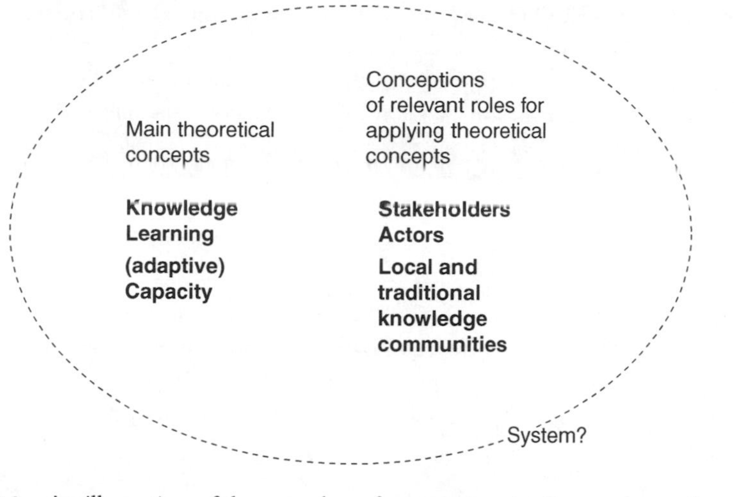

FIGURE 2.3 An illustration of the complex of assumptions in climate change literature (reproduced from Keskitalo and Preston 2019a: 483).

power relations between them are, and how the making and implementation of decisions are effected. "By focusing on the perspective of actors, the understanding of how actors are constrained by structure and co-create structure – something that is a long-running debate in the social sciences . . . – is to some extent lost" (Keskitalo and Preston 2019a).[5]

What does omitting the social knowledge basis result in?

Omitting social science and humanities knowledge thus has a number of more serious implications than what might be commonly assumed.

Among other things, it means that if the social sciences and humanities had been strongly integrated into IPCC and global change research from the start, it is likely that the assumption that "knowledge would be enough" to effect change would never have been emphasised.

It has also been suggested that the exemption of a greater focus on social systems may have resulted in natural science being politicised, instead of the means to address it being the target of discussion. Beck summarises that while there have been attempts to encompass a broader understanding of adaptation in IPCC reports since around 2001, and increasingly more recently, they "are in vain as long as they remain integrated into the linear model of explanation which corresponds to the linear model of expertise" (Beck 2011: 301). The debate comes to focus on a politicisation of the science – the use of science as a political resource – and cherry-picking of scientific uncertainties (Beck 2011), perhaps as there are no actual suggested measures for management expressed to debate. If, for instance, a practical focus on instruments and incentives in relation to social considerations had instead been an integrated part of the study, the debate might instead have concerned these instruments and incentives.

As a result, the focus on the notion that the mainly natural science should by itself motivate decision-making has meant that *this science – rather than the actions and strategies in the governance system that social science may have highlighted in relation to managing issues following from climate change – comes to be the target of discussion*. Beck writes:

> The idea that the IPCC has to deliver the *proof* to compel policy action unduly narrows the scope of debate on adaptation to questions of detection and attribution, thereby marginalizing issues around adaptation and development, distracting from the specific information needs of adaptation policies, the causes of vulnerability, and from the acceptable, grounded solutions available to address them.
>
> *(Beck 2011: 303)*

Thus, Beck suggests, if the focus were no longer placed on the idea that enough science would in itself lead to action, and that accurate prediction would thus in some way be needed from climate change models before action could take place, "[t]he sorts of questions to be addressed would change dramatically" (Beck 2011: 304):

> The emphasis would switch to research that seeks to understand the interaction between climate and society in ways that lead to short-term vulnerabilities (as well as opportunities) in regional contexts. Research would focus more on providing information that is useful for addressing regional and short-term problems – such as health issues and extreme events. These problems also require a better understanding of the socio-economic, political, cultural and environmental processes and paths behind mitigation and adaptation, such as inherent social and economic processes of marginalization and inequality.
>
> *(Beck 2011: 304)*

Similar concerns have been identified for instance with regard to the application of vulnerability assessments, as a part of adaptation research, to natural resource management (Wellstead et al. 2013). Wellstead et al. suggest that such assessments black-box the decision-making mechanisms by disregarding "the complexity of the policy process itself" (Wellstead et al. 2013, para. 14; cf. Palsson et al. 2013: 7). Wellstead et al. write:

> The models upon which they [assessments] draw were developed for other reasons, such as ecosystem impact modeling and studies of community resilience. . . . Although currently in vogue in many geography and natural

> resource management programs, these models are not well suited to the development of feasible policy prescriptions or to the actual practices of policy-making, where the issues of political power, unequal resource distribution, and institutional legacies noted in the case studies are very central concerns.
>
> *(Wellstead et al. 2013, section 2, para. 1)*

As a result, when assessments ask why things are not happening in policy based on such learning or knowledge-focused perspectives and conclude that perhaps participation has not been sufficient (see e.g. Chapter 14 WGII AR5), this answer cannot be assumed to be conclusive. A multitude of research on relevant issues – focusing on the social systems that are relevant to climate change but either not directly discussing climate change or relating to it in other ways than those that are possible to relate to the established framing – is excluded by omission. As seen previously, this may take place through no direct procedural fault of the IPCC assessment system itself, but as a result of the historically developed natural science-focused framing of problems through the linear model and the limited framing of problems and literature that is thereby made relevant.

There is some awareness of this omission, e.g. in IPCC reports; for instance, a 2014 chapter notes that "adaptation literature would benefit by embracing lessons and experiences of mechanisms for enabling institutional change gained in other policy sectors and past policy interventions" (Mimura et al. 2014: 20). There is also an almost general acknowledgement of the role of social science in interdisciplinary research, on a procedural level (e.g. Pedersen 2016). However, these perspectives remain far from integrated, either in the IPCC assessment process through an equal inclusion of social sciences and humanities authors or in the basic research programmes that underlie such conclusions (e.g. Sovacool 2014; Overland and Sovacool 2020) – perhaps because this would entail that at least as much focus be placed on these as on the natural sciences, in a way that is outlined in much of the rest of this volume.

Conclusion

What this chapter has hopefully illustrated is that an inclusion of the social sciences and humanities in all environmental research is absolutely crucial, not only due to any general importance of including multiple disciplines but in order to improve what is currently the limited state of broad programmatic environmental research and assessment. Taken together, this otherwise results in the situation that knowledge in natural science research is not taken up in policy, as it does not sufficiently relate to the actual problems policy-makers face in trying to include climate change (mechanisms that are generally well described in existing social science theories).

Environmental problems – and climate change problems – are thus social problems.

Environmental and climate change issues have often been seen as "wicked" or difficult-to-solve problems (cf. Levin et al. 2012). However, the argumentation in this book is that simply writing off problems as too difficult, as is sometimes done, is overly simplified. A detailed social science analysis of "wicked" problems in their setting – for instance sectoral or other governance frameworks, nested in other governance frameworks – will be able to highlight not only the interests and configurations of actors within this but also what types of incentives can be highlighted to potentially shift actors' interests. As we have seen, positions are given by the present system: Politicians do not necessarily want to go against their constituencies (rather a given in democratic systems), and free trade frameworks set up on the international level are hard to shift as they set the ground rules for each and every actor in the system, so that actors know their positions and what to do to maintain them (see Text Boxes 2.1 and 2.2).

If, however, incentives to maintain positions shift – for instance, if competitiveness becomes (not just seen as, but actually) a question of lowering energy use or of a circular economy, or there are so many storms and insect outbreaks that management needs to shift towards adaptation – this may mean that the positions of actors and frameworks may also shift. It can be a question of external impacts, but also of large governing frameworks shifting due to other reasons, as will be discussed in the following chapters.

It is thus not that change is impossible, but that it has to be understood based on the actual decision-making system and the possibilities afforded within it – not as abstract potential actions in a general sense, but as potentially enabled by the way the specific systems in question are set up.

Using social theory that offers understandings of this is a crucial means to support a better understanding of the social world at large, and of the existing difficulties and variety inherent in trying to understand adaptation and mitigation. To exclude this variation among existing theories from explicit focus, e.g. in assessment reports – beyond, for instance, capacity, knowledge, and learning – is thus to exclude a large part of the accumulated knowledge in the social sciences. However, so far this understanding has not been clearly communicated in large-scale programmatic research in general, much less used as a basis for systematisation in it.[6] To do this, then, the rest of the book will look into how to understand institutions through multiple frameworks; how they are set up, in specific cases; and how such an understanding would shift the understandings that are part of a linear model – the actual roles of knowledge, learning, and power in relation to stakeholders or the social context.

Key points

- The linear model of expertise or scientific knowledge can be seen as having impacted much of how research on environmental problems has been framed and what scientific knowledge has been included.

- Social science knowledge has been largely excluded from the linear model, as it separates science and society. In this light, it has later been seen mainly in relation to communication with stakeholders.
- Social science knowledge, methodology, and theory can provide knowledge about the actual adaptation and mitigation context of actors, including the incentives impacting them.

Study questions

- Why do people following the linear model of scientific knowledge not take their own advice and reject the model, since it has been disproven in research?
- How can the linear model be seen to have impacted how social context is described in the IPCC reports discussed here? (Provide examples.)

Notes

1 That "knowledge is not enough" has long constituted established knowledge in the social sciences and humanities. For instance, environmental psychology and political science would not expect policy-makers to change only on the basis of knowledge – rather, established theories in these disciplines on agenda-setting, framing, and the differences between attitudes and behaviour highlight severe and systematic limitations in the use and application of knowledge (e.g. Kingdon 1984; Schön and Rein 1994, see also footnote 2, Chapter 1, this volume).

2 Sovacool's note on the making of this article, now published in the new journal *Energy Research and Social Science* (which the author was part of developing), was: "This article took six years in the making. Given the focus of its content analysis, an earlier version was initially submitted for publication in four leading energy journals. The author was dismayed that the editors of one top journal in the field of energy technology refused to send the piece out for review on the grounds that it was 'outside the topics' published at the journal, 'fell short' of the journal's 'publication criteria', and would not 'appeal to the majority of its international readership'. Despite our findings, the editors also asserted that their journal 'did not eschew expansion of paper topics'. The editors of a leading energy economics journal did not send the piece out for review on the grounds that 'the subject area did not fall within the journal's scope'. The editor of a leading energy policy journal, after receiving favorable reviews of the article, declined to publish it on the grounds that it, too, was outside the scope of the journal. The editor then proceeded to reject a proposed special issue (much like this one) on the topic. Reviewers for another top energy studies journal argued that 'the paper is not relevant to the scope of this journal'" (Sovacool 2014: 26).

3 Noble et al., in another of the IPCC 2014 chapters, note that "vulnerable groups and institutions often do not have access to the climate impacts science necessary to fully apply top-down impacts-based assessments" (Noble et al. 2014: 852).

4 An example of this, with regard to governance, is that it is noted that "[e]ffective governance is important for the efficient operation of institutions. In general, governance rests on the promotion of democratic and participatory principles as well as on ensuring access to information, knowledge, and networks" (Noble et al. 2014: 849), whereby the "adaptation deficit" is part of a larger "development deficit" (Noble et al. 2014: 839).

5 Similarly, Measham et al. (2011: 889) note that "factors thus far under-acknowledged in the climate adaptation literature . . . include leadership, institutional context and competing planning agendas. These factors can serve as constraints or enabling mechanisms for

achieving climate adaptation depending upon how they are exploited in any given situation". Adger et al. (2011: 757) note that there are "multiple sources of resilience in most systems and hence policy should identify such sources and strengthen capacities to adapt and learn".

6 In relation to the analysis here, Chapters 14–16 in the 2014 WGII AR5, for instance, could have been relevant candidates for being structured to take into account various theoretical understandings of possibilities for change.

Additional readings

Sovacool, B. K. (2014). What are we doing here? Analyzing fifteen years of energy scholarship and proposing a social science research agenda. *Energy Research & Social Science* 1: 1–29.

3

THE ROLE OF THEORY AND CASE STUDIES IN SOCIAL SCIENCE

Means to understanding institutions and contextualising instruments

Introduction

Even if we have not talked about it in these terms, the previous chapters could be seen to have discussed theory. People often assume that theory is less important for real life, as something that is "just theoretical". But theory is not apart from real life; instead, in the best case, it is an abstraction of it. If we do not question and apply theory, we do not know what we are seeing. For all intents and purposes, the linear model of expertise or scientific knowledge works as a theory, including some features and excluding others. The only problem is that it is not a very good theory, and is in fact a disproven one: The world does not work the way the model assumes. And if people do not even recognise that they are in fact applying it as a theory, or how theories work, they have no chance of changing this.

The important issue is then to recognise how theories work and to consciously identify and apply better theories that are closer to real life.

In this volume, in order to be able to conceive of what the relevant components in social science are, a basic understanding is that in social science, theory is explicitly formed and used to understand specific components of society. This is necessary because society is such a complex system (perhaps more so than the natural system as the latter may be more easily defined in relation to universal laws; see Keskitalo et al. 2016). This in itself means that social theories are often built on or derived from multiple cases, but also that multiple theories exist. Results using different established theories will thereby also require an understanding of various assumptions regarding the social system, which will also vary greatly between theories.

This means not only that theory is crucial but also that one has to understand it to use it. It further means that there are limitations to some of the assumptions regarding holistic descriptive possibilities that are inherent in much work on

DOI: 10.4324/9781003043867–3

vulnerability, adaptation, specific systems foci, and resilience, as will be discussed in the following chapters.

The present chapter thereby aims to define social science as a broad field, for students and scholars in potentially other fields but with potential value also for social science students and scholars. The chapter will cover theory as a phenomenon, the need for theory, the applicability and transferability of generalisations to theory, as well as limits to this. It will further discuss particularly the breadth of theories that in some way focus on what have been among the less focused-on features in the linear model of scientific knowledge: institutional perspectives. The chapter will also highlight the role of case studies, as studies that may allow a better understanding of the organisations, mechanisms, instruments, and actors that are relevant to adaptation and mitigation, and it does so even in cases in which climate change has not been the focus of the study.

Theories as well as case studies that in some way highlight institutions as more stable features of social life, but also what may be required to change them, can thereby serve to understand some of the features that have been under-analysed in conceptions that apply the linear model of scientific knowledge – and can be used to better understand society as it is. In total, the chapter thus problematises the assumptions in the linear model and aims to substitute them with theories of how social change actually takes place, as a foundation for empirically, reality-based study and any attempt at effecting change.

A focus on the roles of theory, case studies, and institutional perspectives

This chapter takes its point of departure in the fact that social science theory largely holds the same role as results may do in natural science. In social science, particularly in qualitative studies or case studies, *theory* is the body of knowledge that is generalised to and modified by results in the specific case study. This is because cases may vary so much that the case study cannot be generalised; however, it can show more general mechanisms or impacting factors that add to or reinforce those discussed in literature (e.g. Flyvbjerg 2006).

This role of generalisation to theory, particularly in qualitative studies, has seldom been recognised in climate change literature, where concern has instead been expressed about studies being case studies to such a high extent.[1] Reviewing the much larger body of literature, however, and going beyond a focus on knowledge, learning, limits, barriers, and capacities (as discussed in the previous chapter), opens up to a much wider and already well-studied plethora of relevant theory and theoretical frameworks. These have already been applied, for instance, in case studies that can be directly relevant to the climate change case. Case studies of specific organisations, mechanisms, or levels – even if they do not discuss climate change – can say something about an organisation, mechanism, or level that plays a role in adapting to or mitigating climate change. And if this organisation, for instance, can be understood, this may also say something about how it can be incentivised to act on adaptation or mitigation.

Theories are thus important because they are what can be generalised to based on a case, and cases can be important as they highlight features of an organisation, mechanism, or other entity or actor that is relevant to adaptation or mitigation. A case is not just a singular case but rather a "case of" some broader phenomenon that can then be compared to; this is the point of case studies (e.g. Flyvbjerg 2006).

It is thereby important to understand what a theory and a case are, and also why they are selected. In this regard, not being able to cover all potentially applicable theory here, this chapter focuses on the omission of the more *structural* or "system" context of social life discussed in the previous chapter. This does not mean that theory focusing on the actor level, or individual stakeholders, is not needed; but as theories often used in adaptation highlight actors, there is a need to also include structure (cf. Potter and Tilzey 2005). This book thus does not (at all) aim to replace all agency-focused theory with more structure-focused theory, but rather to provide the theoretical additions that are needed to not only focus on actors. Instead, the volume, in this and following chapters, focuses on cases that can highlight different actors, organisations, mechanisms, or instruments that are relevant to environmental change (taking the example of adaptation).

Another selection in this as well as following chapters entails applying a study and theoretical approach that highlights *real-life* situations and the role of power. The book thereby aims to distinguish itself from perspectives that do not focus on actual real-life analyses. This speaks to another consideration in understanding theory: to also understand that theory can be developed for a large number of different reasons, some of which do not focus primarily on understanding real life. Theory can also involve, for instance, thought experiments, or normative or ideal theory, which may describe how the world should be rather than what it is. Different from these types of theory, the theories discussed in this book are the kind that focus on analysing and understanding real-life situations, applied to real-life cases. The argument is that it is theories focused around understanding real-life situations that are applicable in identifying real-life institutional features, as well as which potential instruments could be designed to impact them. In this, the volume thus differentiates itself from not only more speculative approaches such as scenario-building, as mentioned earlier, but also from other theories and studies that are speculative, normative, or ideal (an example of which will be discussed in Chapter 5).

As a result of this focus on conceptions applicable on a system or structural level as well as in real-life situations, this chapter takes its point of departure in a focus on *institutions*. The argument here is that understanding existing institutions and their embedded incentives and constituencies is the best way to understand the potential for developing adaptation and mitigation (or, more broadly, action in relation to the environment) among them; as most of the organisations, for instance, that are called upon to adapt to and mitigate climate change already exist and are historically developed to speak to other aims. This means that institutional features may entail lock-in or limitations with regard to how much these organisations can orient towards new aims and still maintain existing aims and constituencies, unless other organisations with which they are interlinked also shift simultaneously – something that is very hard to do, as will be discussed in this chapter. Institutional theory also covers not only the persistence of institutions but also how they can be changed. The issue of change, one that is implicitly in focus in the linear model of scientific knowledge and in many related studies, is thus also in focus in institutional theory. However, in institutional theory change is not assumed but is rather made a focus of study. Rather than assuming that change will take place, the question is: What mechanisms may support change?

In principle, a focus on institutions, and on institutional components of case studies, may thus illustrate why there are limitations to the direct implementation of knowledge or learning. This focus may also highlight that a key consideration regarding implementing something like climate change adaptation and mitigation is *understanding the systems in which it is to be implemented* in order to design legislation, policies, or other instruments to support change. Given that the focus of the chapter – and the book as a whole – is on understanding the implications of the fact that the world we live in is one made up of institutions, in a broad sense, the focus will be on illustrating this fact, drawing on broad and varying ranges of theories as well as cases from various fields. However, this chapter, indeed the book as a whole, does not aim to be a textbook in institutionalism. Instead, it draws on varying theories which specialists in any of them will recognise as indeed sketchily described, with the aim of illustrating relatively broad agreement across different social science fields that institutionalised elements matter (even if they are called different things in other bodies of theory). The term "institution" will to that aim be used here not to refer to any one selection of theory but rather to the notion that larger, or perhaps all, aspects of the social world are institutionalised (Thornton et al. 2012) – that is, they cannot be assumed to change with, for instance, improved knowledge only.[2] In this way, the main intended claim is that actors or stakeholders are in some way formed, constrained, and also enabled with regard to different ways of action, involving specific features that may vary between different social systems. As a result, while the description that follows draws to some extent more on political science literature – as it includes a focus on the higher levels that are relevant to add to a focus on the individual stakeholder or local level in climate change literature – the intention is not to exempt other perspectives that are relevant to this type of logic (Thornton et al. 2012). Multiple theoretical perspectives, even other than those mentioned in this book – may be relevant to what are here seen as broadly institutional understandings.[3]

Why is climate change a question of understanding theory, cases, and institutional dynamics?

The social world is incredibly complex. For any one situation you study, you have to identify what you are looking at in relation to what might make a difference or be most important in order to ensure that you get at least the necessary information on these parameters.

This is similar to a hypothesis in the natural sciences, asking for instance: What do you think matters? As the social world is so complex, it is also not enough to say something like "I think language, how people speak about things, is what matters". What do you mean by language? Is language only what is said, or is it also what is communicated in some other way (e.g. body language)? Or do you look at how people are able to speak about certain things; that is, beyond language itself? And given all this, if you attempt to study a decision-making situation by studying "language", how do you ascertain that you are actually getting at all the parameters that are relevant to "language"? And if you only study "language", what can this actually say about the decision-making situation? (Lakoff and Johnson 1980).

For these reasons involving the massive complexity in social study, theories exist. In the social sciences, as noted earlier, theoretical frameworks are not simply "theory" in some meaning of being removed from reality (Moran 2010; Gillard et al. 2016). Rather, they are often condensed reality, in the meaning of defining what mechanisms are assumed and investigated as to whether they matter in relation to the specific phenomenon or focus being targeted.

As an example: Many people who have perhaps begun by wanting to study "language" have defined different understandings of what "language" is and how you study it. Some, for instance, may have decided that the most important aspect is what knowledge people express and how others can learn it, while others may have decided that the situations that make specific statements possible to express are the most important. These would constitute very different understandings of how to study "language" (Lakoff and Johnson 1980). Different theories would then also be chosen concerning what part of the phenomenon "language" you want to study, without necessarily invalidating other theories that look at other parts of the phenomenon (although some theories may take such different points of departure that they cannot be combined, as will be discussed).

This complexity of the phenomenon, which of its characteristics are chosen to study, and the complexity of theory alike, is the underlying foundation in social studies. For these reasons, publications are typically required to contain not just Introduction – Methods and materials – Results – Discussion (IMRAD), as is the norm in natural sciences, but also a Theory section. This is also one reason why social science studies often need more space (words) to describe their results: Results have to be qualified in relation to theory, and theory has to be described so precisely as to enable readers to understand what was really studied, as different authors may highlight different parts of specific theories in order to study specific aspects, for instance of their case.

Theory as encapsulated reality: that which you generalise to in qualitative studies

Theory is also the way in which results from what may be – and often are – very different cases are made generalisable (Ruddin 2006). Given that, among other things, the social world has been formed historically over time, resulting in widely different situations at different local, regional, and national levels, it cannot be assumed that the situation per se will be the same in different cases; rather, it can be assumed that it will differ. For these reasons, qualitative studies or case studies – often used in, e.g., adaptation research – are thereby not some type of "lower quality" studies. This is a misconception and, following the previous chapters, not surprisingly potentially related to an often more quantitative focus in the dominating natural sciences and the influence this has had on social science as well (Flyvbjerg 2006). *Instead, to understand and correctly deal with the great variation in the social system across localities, nations, and levels, qualitative studies are often the most appropriate tool.* Qualitative studies are often seen as targeting "understanding" – why people do something or why an organisation has a certain design, for instance – while quantitative studies are often seen as targeting explaining: how frequent some already identified phenomenon is, what is the extent of already identified parameters (e.g. Hollis and Smith 1991). This means that in explanatory studies one does need understanding in order to know what one is studying; qualitative studies are a necessary part of this.

But qualitative studies do not work the same as quantitative studies: The role of theory is not just important but paramount to the type of "explanation" qualitative studies offer. This is because, while various cases viewed at different national, regional, and local levels may differ, what could be assumed is that the mechanisms for something may be similar. For instance, language (in any specific theoretical understanding) may gain influence in the same ways but depend on specific factors in the different cases. As a result of this, in qualitative studies, generalisation is made not to a population (as the interviewees, for instance, have not been selected to be representative of a full population) but *to theory*. Interviewees may have been selected, for instance, in relation to different theoretical parameters, or to be as different as possible, so that any commonalities between them concerning parameters in the theoretical framework will then be able to say something in relation to theory. If very different actors (or whatever unit is investigated) act in relation to what is assumed in theory, this says something about the theory (although it can be disproved in other cultural circumstances or contexts) (cf. Sørensen and Torfin 2014).

Theory is thereby that which you generalise to in qualitative studies. And by doing so, you also continuously review and potentially suggest modifications to theory. This means that the best case studies are never single cases only, unconnected to theory: They are cases that are related to established theory, that describe why mechanisms work in the same or different way in the cases, and that can thereby be generalised to theory. In the best, most clear cases of such an application, the cases may also result in suggestions for changes or modifications to theory, so that the way mechanisms work in different cases can be better understood (Flyvbjerg 2006; Ruddin 2006). This is "explanation", but not in terms of numbers or statistics but of how things work, and it first and foremost requires an understanding of what the specific situations are: the case and the institutional context.

A case study is not only a case study

Following from the previous section, a case study is thus typically not only a case study. Instead, the fact that there are many existing theories and studies of the organisations, mechanisms, and actors relevant to climate change adaptation and mitigation means that we already know a great deal about adaptation and mitigation.

Even if adaptation and mitigation per se are relatively recent policy issues in that they have only gained traction in the latest generation or so, placing the focus on theory as well as on the organisations, mechanisms, and actors that are studied in social science studies means that we are "far from starting from scratch with regard to understanding the system into which it needs to be integrated to gain traction" (Keskitalo and Preston 2019a: 478).

This is because, given the role of theory as well as case studies in social sciences, there are already studies on the types of *mechanisms* (or organisations, or actors) that are relevant to adaptation and mitigation. And given that adaptation and mitigation need to be undertaken by large institutions and organisations such as different sectors, the state/national level, and subregional and local levels in different countries on which there is already a large body of research – on their working styles, different instruments that have been implemented in different contexts, organisational motivations, and the like – the knowledge is largely already there with regard to mechanisms that may become relevant in different cases (e.g. Jordan et al. 2013; Finnemore and Toope 2001).

This highlights a crucial issue that is not typically emphasised in climate change research: the fact that in order to understand the possibilities to *respond to and limit* climate change, understanding climate change *itself* may be less important than understanding the *decision-making system* at different levels and degrees of formality, within which responses to climate change would take place and need to be embedded (Keskitalo and Preston 2019a). We already know a great deal about the organisations, actors, and mechanisms that constitute "barriers" or "opportunities", and they have to be conceptualised much more specifically than this.

The "meta" of social science thinking: theory and case as a way to understand what is happening

Social science thinking thereby does not mean that only theory is important, or that only the case is important (as has often been the assumption in criticism of studies for "only being case studies"). Rather, both of these are means to an end in the type of real-life studies and approaches in focus here. It is thus not only "system description" per se that is crucial but also an understanding of the mechanisms: the "*how*" of how something happens. For instance: What mechanisms influence decisions made where and when, and in what way? Different theories cover this in different ways, and different cases can be used to refine theory. Theories can also be used complementarily to understand different parts of a phenomenon, such as how something happens, and cases can be used to understand how different actors or organisations respond, are limited in the ways they can respond, or the like.

This type of "meta" thinking, which is in many ways inherent to much of social science, has seldom been recognised in natural science assumptions (and is absent from any social conception related to the linear model of scientific knowledge). Social science is not always about "system description" in a way that can define a system's boundaries, limitations, or barriers – in fact, defining such things is not always possible (e.g. Olsson et al. 2015). Instead, much of social science is about understanding both theory itself and how it can be applied (and what the limits of this entail), as well as cases and their applicability (What is something a case of, and what broader category is it selected to illustrate?).

Almost needless to say, applying such an understanding, focused on theory and different categories of actors, for instance in different countries, might offer radically different understandings of capacities for social action than do, for instance, conceptions that focus on capacity or barriers. Understanding theories or cases in this type of context, however, requires a good understanding of both the degree of complementarity of different theories and what parts of different phenomena they focus on. It also requires a good understanding – or problematisation – of what the "system" is like and how results gained in one case may differ from those gained in another. We cannot just go to a body of literature and take something out for implementation or application because it exists in that one body of literature: Neither any one theory nor any one instrument (for instance a specific regulation, strategy, or the like) can be assumed to provide all the answers. If we are looking at a theory, we have to consider what its main characteristics would entail for the situation we are looking at and also whether there are other or related bodies of theory that can support our understanding. And if we are looking at an instrument, for instance a strategy, as much adaptation research does, and trying to assess whether it will be the most relevant tool for achieving change, what we need to do is at best – given the complexity of the social world – consider where and when such instruments have been implemented, in what context, whether the relevant actors, instruments, or organisational context have changed since then or are different in this case, and whether the mechanisms that worked for something in the past or elsewhere are likely to work for something else right now (cf. Victor 2015; Lubatkin et al. 2005).

Thus, for instance, specific adaptation or mitigation-relevant strategies, actions, or instruments that have been identified may work differently from ones applied earlier, or differently in one context compared to another, for instance as situations have changed or because these actions or instruments imply different actors or processes. Many environmental governance instruments in fact embody assumptions derived from the context for which they were originally developed (e.g. Lubatkin et al. 2005). This may explain why, for instance, environmental governance instruments that were identified in a market context to be undertaken by companies may not be working for the purposes of public actors, who often have to respond to multiple aims rather than mainly market aims (Keskitalo and Andersson 2017). In a similar vein, it has also been shown that more voluntary instruments only work in highly specific cases, for instance where actors are already more broadly incentivised to apply them (e.g. Alberini and Segerson 2002).

The role of context dependence: the institutional context

Context dependence is thereby important to understand. The fact that there is already a great deal of knowledge in the literature does not imply that something can be simply taken from somewhere and be applied elsewhere: The context, actors, instrument, or anything else in the setting or application may vary, meaning that the results of any implementation (for instance, of mitigation or adaptation measures) will not necessarily be the same. Any implementation of measures will also not only involve the instrument or similar (the thing to be implemented or that is intended to evoke change), but the process as well. Something that is implemented in a specific way may lead to other results than the same thing implemented in a different way. This would be because the process may differ: If something is done in a way that angers people rather than appeals to them, it might not be accepted even if it has the potential to be beneficial. For these reasons, numerous mechanisms matter, and the thing to be implemented cannot be seen as apart or aside from the implementation context (e.g. Moran 2010; Gillard et al. 2016). The theoretical frameworks that, for instance, describe what works where (such as in the case of the policy process or implementation) often highlight specific requirements for specific processes. It is then important to understand what these are as well as what they might mean in the specific case, if one looks at possibilities to support change. It is also important to understand whether there are other complementary theories that can be applied to add to the knowledge.

Context dependence, then, is not just a fancy term to say that something is difficult; it is a real consideration that must be understood in order to apply lessons from theories or case studies. What makes this theory relevant here? Is what we are looking at a case of some other type of phenomenon that has already been studied, and how similar to those cases may it be? What are the defining characteristics of the different types of cases, in different regions or countries, and to what extent are they the same?

Institutions, then, are important as they highlight the underlying and to some extent more stable nature or fabric of society. They thereby highlight not only change – a crucial part of understanding implications of climate change – but also why change does not take place.

For these reasons, the following section will focus on what could be seen as the major components in understanding institutions: the way institutions imply a limitation to change (and thereby also to knowledge resulting in learning or change). While acknowledging that not all theory is transferrable or generalisable, but that it depends on context, as noted earlier, the aim in this section is to stick to very broad foundational understandings of change that could be seen to bridge several bodies of literature and that perhaps most social scientists would recognise parts of (if under other names) (Thornton et al. 2012). Later sections will then problematise what can be said beyond this type of generality, for instance with regard to implementing an understanding of these principles for climate change adaptation or mitigation.

What does a focus on institutions imply?

At its most general, most of social science could be said to be about change, and how it is enabled or delimited. Institutional perspectives are made relevant here as they relate to a wide variety of structures on different levels. Institutional change is discussed in a large number of both related and separate bodies of literature, for instance in relation to incremental and catalytic change treated in historical institutionalism and other fields of study (e.g. Kingston and Caballero 2009; Streeck and Thelen 2005; Mahoney and Thelen 2010). However, something like discourse analysis – which is seldom seen as institutionalist – also highlights the way present structures are maintained or able to shift (e.g. Foucault 1974), as do historical-technical approaches like some science technology studies (e.g. Meadowcroft 2009).

This wide variety speaks to different aspects regarding the wide range of features and their malleability that are typically included in definitions of "institution", even when applied in more delimited areas of study. In the field of institutional change, Oran Young defined institutions in the international *Institutional Dimensions of Global Environmental Change* programme science plan at the most general level as "systems of rules, decision-making procedures, and programs that give rise to social practices, assign roles to the participants in these practices, and guide interactions among the occupants of the relevant roles" (Young 1999: 27). Institutions are thus in a very broad sense rules for human interaction, or "rules of the game" (as suggested by Nobel Prize winner Douglass North 1990: 3, cf. 1994).

What this fundamentally means that any action is always bounded by institutions. Institutions are sometimes those one is aware of, and sometimes those that are so naturalised that their rules are quite simply seen as given. What choices one makes are thus inherently related to the institutional structure within which one acts. The institutional framework provides a structure for social and economic interaction (defines the opportunity set) by imposing constraints through norms and regulations (North 1990, 1994). Institutions are the basis for both formal and informal governance systems. The formal system, which may be what we most often think of as "governance", is built up of international, state, regional, and local governments or bodies, but also institutions such as the market (Piattoni 2010; Hooghe et al. 2019; Hooghe and Marks 2001; Marks and Hooghe 2004). However, it also comprises and is comprised by the ways these are developed and maintained – by sectorial organisations across scales and by numerous supporting subsystems, such as standard-setting organisations and practices (Burritt and Welch 1997; Talbot 2000). In a way we may not always think about, all these formal institutions are also supported by numerous informal institutions: practices and assumptions reflecting "this is just the way we do it". But "just the way we do it" may actually be even more determining than formal and directly identifiable institutions, as it may indicate what are sometimes very strong and determining social rules, or more habitual and thereby sometimes almost given actions in the specific social setting. "What we do" is also always necessarily nested within the formal and informal power structures that enable some actions and delimit others (Mahoney and Thelen 2010). This means that "institutions" are not just things that exist on higher levels or in formal governance but rather things that exist and govern actions everywhere, even at the local, group, or individual level (even if they are typically and traditionally addressed by other terms there) (Thornton et al. 2012).

This brings us to the topic of change: As institutions are defined as related to rules and thereby relatively enduring practices, they are by definition related more to being resistant to change than amenable to it. Institutions in various ways target what some might call structural or system characteristics, but in a way this does not remove them from individual – or what some might call bottom-up – action within them. As Mahoney and Thelen note: "What institutions do is to stabilize expectations (among other ways, by providing information about the probable behavior of others)" (Mahoney and Thelen 2010: 10). Institutions may thereby provide the basis for interaction on multiple levels. Formal institutions like laws are cases in point: As their very intention is to provide a kind of baseline so that treatment is the same for all, they are not meant to be easily changed (Edelman and Talesh 2011; Keskitalo and Pettersson 2016). They also often (but not always) encase norms that are agreed on in a society and that relate to acceptable behaviour or assumptions involving interaction. However, this type of focus on stability and not being easily changed is not restricted to formal institutions – rather, informal institutions can be as demanding, and sometimes more so, as we may not be able to identify and question them the same way as we can formal institutions (cf. Foucault 1974).[4]

Change is not a given

While this does not mean change is not possible, it does mean that "transformation" (e.g. Park et al. 2012) cannot be assumed to take place in any given case. There is relative consensus across the social sciences regarding this type of acknowledgement of limitations to change. Change cannot be seen as given, but will depend on how the forces that sustain institutions are influenced: how issues of change are made urgent or not, including how they are framed and expressed, and by whom; what other issues or requirements these changes compete with; and the like (Manicas 2006; Béland 2005; Harty 2005; MacCormick 2007).

Change is thereby seen not as directly related to any external general requirement or externally conceived "rational" course of action, but rather it depends on how any such requirements or others are made relevant or even urgent to the institutional structure, and notably to existing power arrangements (North 1994; Geels 2011). Depending on how changes are made urgent or relevant in different ways, they can be sudden (for instance, as a result of power shifts) or gradual (for instance, as preferences and opinions change) or take place as the basis for specific practices is eroded as a result of changes in other systems or subsystems. They can also overlap or not take place at all (e.g. Gingrich 2015).

Also at this level of generality, a focus on institutions – as broad types of structures constraining change identified in much social theory – thus disproves the understanding in the linear model of scientific knowledge: Knowledge as it is prioritised by and developed in one specific context cannot be assumed to be prioritised in another. It can also not be assumed to be learned or implemented in any automatic way, and it cannot be assumed that bottom-up actors would necessarily be able to act in relation to larger structures; while they may, this is not a given but rather formed by the types of contexts described earlier.

Both formal and informal institutions of multiple types interact, making change complex

This role of change, largely in relation to encased power dynamics, also highlights the relation between formal and informal rules or governance or steering. What are now formal and explicated rules have often evolved ad hoc to an extent of formal agreement (for instance, as they may be in contested areas where a need for formal agreement was developed). By processes of contestation in some form, formal rules are continuously confronted with evolving informal rules in their areas, and informal rules may be challenged by other ways of being or acting.[5] As a result, it is crucial to comprehend existing power structures as well as political contestations within the system and how they are managed. Contestations, or external demands, do not necessarily result in things simply shifting. If partially successful they may, for instance, result in competing institutionalisations that supplant rather than replace existing rules. Thereby, they may create competing systems that make it possible for actors to proceed without changing logics or strong existing decision paths in the system (Olsen 2009).

As a result, knowledge of a "system" cannot be defined without knowledge of the multiple types of both formal and informal institutions that exist. This means that assumptions that focus on defining "system" limits, particularly for smaller "systems", and attempt to determine decision paths based on, for instance, more cursory studies targeting them are most likely too delimited (e.g. Olsson et al. 2015). Instead, an understanding of the mechanisms that may support change may most likely also need to relate to what can be made relevant to actors in any case, depending not only on directly identifiable "objective" features that are easy to distinguish but possibly also on ranges of features from culturally based assumptions to higher-level regulation (far from, for instance, local "systems") (Keskitalo et al. 2016).

Developing "change" is thereby not any one clear path, and it is also possible to avoid, renegotiate, or go back on, or develop partially or in limited form. This complex nature of possible change in institutional systems means that transformational, catalytic, radical, or "faster or larger" change is not always possible, and also that larger changes may also not always be more effective or indeed take a shorter time than a number of more incremental changes that build on each other. Not only may incremental changes offer actors specific benefits, but sequences of incremental changes could lead to faster change than single radical changes, as the latter may be more likely to be resisted or implemented in a way that does not support the original aims (Marsden et al. 2014).

Change is not just incremental or radical

It is also not as easy as saying changes are either incremental or radical – slow and gradual or fast and catalytic. Instead, the two different types of changes may support each other or work in sequence, for instance as thresholds are reached whereby a change becomes accepted or issue severity is recognised (discussed in literature as, e.g., "punctuated equilibria", e.g. Baumgartner and Jones 1993; cf. Sapotichne et al. 2013).

Fast change is also – in the literature since Lindblom (1959) – well recognised as not always positive per se. Incremental (slow and gradual) change ensures a preservation of results of the past and structures that have co-evolved and are workable to decision-makers. As institutions are both formal and informal, a "workable" institutional system often reflects that decision-makers are able to act within it or are even naturalised into it without necessarily having to clarify guiding values and analyse how to reach specific aims. In fact, a great deal of policy analysis literature has highlighted that much decision-making – contrary to a rationalist assumption – works by people not fully understanding or confronting how decision-making structures are set up: by "muddling through" (Lindblom 1959; Cohen et al. 1972; March and Olsen 1996). This is because structures are generally so complex – made up of both formal and informal, changing, components – that people need to learn to operate in them by practice. This is not necessarily negative – mostly, it is a fact that needs to be related to. However, it may mean that there remain several ways to get to specific outcomes even under conflict, while retaining room for manoeuvre (e.g. Lindblom 1959; Marsden et al. 2014). It also means that there are multiple issues to take into account concerning how systems can be shifted.

While this means that systems are not necessarily easy to steer, it does not mean that steering is impossible; it goes on every day and provides for changes. However, it does mean that change is not unidirectional or will necessarily be able to supplant existing structures. It also means that what one has to pay perhaps the most attention to are the encased power dynamics in a system and how different changes might impact them: Who benefits from certain changes, and who loses, and over what time and at what levels? What are the possibilities, in these systems, that other actors may act in supportive ways, or be able to change routines, so that specific changes are meaningful for actors to undertake? And what are the possibilities that a change at one level under one specific understanding will be possible to transplant to other parts of a system with other prerequisites?

Thus, agenda-setting theories on the policy process (e.g. Kingdon 1995; cf. Zahariadis 1997) generally emphasise the competition between issues to get on the policy agenda: An issue is only raised from a background condition as something that demands real policy attention due to, for instance, events or changes in political or policy context that provide those arguing the issue with room to forward work on it and extend it. This means that the configuration of actors or interests, and how it has been enabled, becomes important. It is also crucial to note that how determining an institution may be thus does not necessarily involve whether it is formal or informal – although formal rules may have the most easily identifiable impact on higher levels – but rather how embedded it is. Rules or institutions – or discourses – that are well embedded, with high implicit or explicit (often both) agreement among powerful actors, may be the most difficult to change (e.g. Kingston and Caballero 2009; Foucault 1974): In fact, the linear model of scientific knowledge could be regarded as an element of such an embedded assumption, which is seldom explicitly recognised by those acting in relation to it, and which is thereby difficult to change. These types of structures are often developed in relation to historical configurations of what actors are powerful, and they are able to dictate specific logics or ways of seeing or acting in relation to specific issues.

Change in relation to paths or logics

Understanding institutional change and the processes that support it may thus also necessitate a longer-term historical – and future – perspective, to capture "the organizational processes through which compromises and victories in political battles are 'frozen' into institutions, sustaining a lasting legacy" (Olsen 2009: 24–25). In this context, historical institutionalism or path-dependency theoretical frameworks, among others, have been used to explain why change is difficult (e.g. North 1994). Different types of perspectives, which in some kind of way highlight that change is related to "paths" of what is workable and what the costs of change are, have been highlighted in different ways by various authors, ranging from policy studies to science technology studies and economic history, drawing on studies in relation to institutional, technical, and socio-technical systems and highlighting amongst other things the economic/investment, social, cultural, and knowledge costs of change (e.g. David 1986; Kemp and Soete 1990; cf. Meadowcroft 2009). Among other things, this type of literature illustrates that decisions made in the past affect what choices are available in the future, as well as the relative benefits actors see in routines – doing things the same way, thereby contributing to self-reinforcing systems and raising the system costs of change (e.g. Pierson 2000). System and actor costs of change are also raised by the notion that subsystems may not be possible to change independently (e.g. Liebowitz and Margolis 1995).

For the purposes in this volume, a general understanding of issues around path dependency can be seen to focus on understanding the need for clear and potentially even demanding, system-wide, incentives for actors to change development paths, as system and subsystems that are internally dependent on each other create "lock-in" effects that make change to established routines difficult

(Greener 2005). Again, this does not mean change is impossible (nor does it mean we should subscribe to a "hard path dependency" perspective).[6] What it does mean is that historical contingencies and the ways institutions and the interests and subsystems they support are set up in the present cannot be assumed to be the most effective ones. As adaptation literature has also noted, we are not perfectly adapted even to present conditions (McDonald 2011; Smith et al. 2009; Preston et al. 2011). This is the case particularly as institutions have over time been oriented to those goals that "have been the most instrumental over time in providing a benefit to those invested in it, and which have come to be embedded into being, thought, and action" (Keskitalo et al. 2019: 8). This often means that they have targeted, for instance, national economic goals relevant to the state and strong interest groups as well as practices, rather than goals that are often conceived of as external to the system, such as environmental ones (e.g. Van den Bergh 2013).

Thus, based on the role of established institutions and interests in influencing the present as well as guiding their future decisions, it is what could be seen as the *logics and interests that are embedded into institutions* over time that largely determine responses not only to climate change but to all kinds of change – posing challenges for those who wish to promote, for instance, responses to environmental change. What an institutional understanding of possibilities for change thereby mainly implies is that one needs to focus on understanding the existing environment of institutions with their embedded contexts and assumptions, and not least the driving forces and the complex, interlocking motivations and incentives. While it is a relatively general conclusion, it nevertheless runs contrary to assumptions in much work on climate change that has largely omitted a focus on the detailed mechanisms in decision-making bodies (having "black-boxed" the policy system, as Wellstead et al. 2013 see it), and thereby the role of historically developed processes and the present power configurations. This provides an important perspective contrary and adding to, for example, climate change literature, which has largely assumed that knowledge of the scope and severity of potential change will be sufficient for decision-makers to change the institutions where they work (cf. IPCC 2007, 2014) or which has based its comprehension of possible future social change more on potential scenarios than on a solid, empirical understanding of the past (e.g. Kates 1985; cf. Walz et al. 2007).

To avoid treating the system like a blank page or black box into which the climate change issue is inserted, a broad institutionally based understanding may thus help us understand not only how climate change affects the system, but also how any new issue entering it may be influenced and possible to influence. In this sense, the suggestion would be that an understanding of responses to both climate change and any other new emerging issues largely needs to be based on the ways the existing system already selects and biases particular decision paths – and non-decision paths. This would mean, for instance, that adaptation paths can be chosen that resonate with different decision-makers and within different sectors, even without explicitly dealing with climate as such or making it the focus of study.

In sum, a way to conceive of limitations to the implementation of knowledge or actions – not only concerning climate change but other issues as well – is thus to review the ways in which change has previously been effected (or not) in the multilevel social system case (Meadowcroft 2011) and to understand the incentive and power structures under which actors act. Lock-ins, path dependency, the role of institutions, slow or fast change, and cost-benefit (taking into account the numerous costs and benefits relevant to actors, not only economic ones) constitute different ways to conceive of the notion that society is never an unwritten page: What can be done, while far from being entirely determined, is at least preconditioned by what exists. This is why, contrary to work in mainstream climate change research and based on the understanding sketched previously, understanding the institutional trajectories of the past into the present and the way present institutions govern action may constitute a more potent tool than established scenario research for comprehending what future change may be likely and what may constrain or limit it.

Can institutional change be influenced – and "solutions" found?

The previous section illustrated some of the general tenets around thinking about institutions and change, and how change is circumscribed. This type of knowledge, under this generality, may be established sufficiently to, for instance, constitute a basis for thinking about change in broad research programmes or assessments – a step up from, for instance, barriers/opportunities thinking. However, this depth of understanding of theory is seldom reflected in the climate change debate: Far from a notion that social science would not be "solution"-oriented (Watts 2017), it may be that the types of "solutions" – in the form of mechanisms that play a role in what happens in a situation – that can be identified from an understanding of what different types of theory may contribute have not been recognised. This is because, for instance, the role of theory not least in relation to case studies – even outside the climate change field, on cases, organisations, or levels relevant to adaptation and mitigation – has seldom been recognised. In addition, when it comes to identifying "solutions", it has likewise seldom been explicitly recognised that social studies may target different levels of specification: more overarching theoretical ones as well as highly specific ones that may be subject only to highly specialised theory, as well as studies of instruments ("solutions" that can be applied) but that may be extremely context-dependent.

While overarching explanations, for instance of how change may be delimited, can be identified broadly in social theory – and themselves add to the limited social understanding manifested in things like IPCC reports – getting to "solutions" and "what to do" needs social study: What is the case and context, what are the intended effects, what instruments are considered to be applied, and so on? That is, a focus on social theory does not limit the need for social research, but social theory based in empirical analysis provides a reality-based framing from which to identify the broad concerns and instruments that can be applied as "solutions", as well as to develop extended and specific study. We must also be aware that social study is not a magic bullet: As soon as we may start thinking we should find any general "solutions" (e.g. Watts 2017), of a type more specific than on the level of mechanisms that can be identified in theories, the devil is in the details. Instead, the types of "solutions" as to how in detail to steer policy entail a much more difficult level than that of general theory, as the answers will differ between, for instance, states and levels.

So, as soon as we stop focusing solely on understandings based on theory that have been applied in highly different contexts and start focusing, for instance, on *the instruments or practical policy tools* themselves, in line with what is discussed earlier, it is likely that the instruments or tools may not be directly transferrable. In fact, the variation in governance systems that one may try to steer by applying policy instruments is enormous and has not always been recognised (e.g. Lubatkin et al. 2005): In attempting to find more "generalisable" principles for what works "beyond case studies", it may be that the baby has been thrown out with the bath water and the huge role of social science in illustrating issues particularly in relation to the role of context dependence has been misunderstood.

Discussing policy instruments is thereby relevant, as it can provide both an illustration of limitations to social science knowledge – which are similar to the large applicability of more general social sciences knowledge, less discussed in climate change literature – and of the more specific difficulties involved with gaining traction on climate change adaptation or mitigation. It is a type of study that is often related to an institutional perspective; however, it is also a class of study in its own. What much of the literature does, however, is to highlight the institutional context in the way we have understood it here. There is a large body of literature, particularly in the policy sciences, on "policy instruments", which are the regulative, economic, and information measures that, for instance, a state may use to effect change. In other words, the idea is that you can effect, motivate, or incentivise change by legislation or taxes (or other economic means) or by providing information or other voluntary means (e.g. Gunningham 2009; Rosen-Zvi 2011). These instruments are typically not seen as "catch alls" but as instruments that have to be adjusted to context and goals. Over the last generation or so, all these types of instruments have been increasingly added to by voluntary and informational means such as certification or labelling (Jordan et al. 2013; Lenschow 2014). These voluntary instruments, different in that they were often implemented by private actors such as companies rather than by states, were long regarded with great hope: that people would voluntarily, and through their market power as consumers, "choose" options that then outcompeted less environmentally sustainable ones (e.g. Alberini and Segerson 2002). They thus carried with themselves specific assumptions regarding how people acted and were thereby also designated to specific solutions or institutional contexts.[7]

Understand the extent to which policy instruments can be applied as "solutions"

To understand instruments in context, then, not only is there a need to understand how variation in contexts, particularly for the same instrument or group of instruments in comparative cases, may influence implementation (Jordan et al. 2013; Finnemore and Toope 2001). There is also a need to understand this through an institutional lens concerning how different, for instance, regulative assumptions in different systems, and how they are steered or what characteristics they hold, may lead to the application or non-application of instruments (Lodge 2014; Jordan et al. 2013). Firstly, in line with what has previously been noted, it cannot be assumed that actors will, for instance, follow a knowledge-based "rational" approach by which they pay attention to all aspects of a new problem – instead, it must be recognised that the contextual and institutional characteristics of actors and situations are often what determine the decision-making method and orientation that are possible. Marsden et al. note:

> [W]hat is deemed a rational approach to a problem depends on the circumstances of the decision-making process. . . . [W]hen a group of actors with clear and consensual goals meet in an environment characterized by simple interactions, perfect information, and plenty of time to make decisions, there are good conditions to apply the comprehensive method. However, as these conditions mutate towards greater complexity, the comprehensive approach is no longer suitable.
>
> *(Marsden et al. 2014: 620)*

Secondly, in many cases, these types of considerations that can be drawn from theory based on real-life situations were not clearly applied to implementation or research on instruments. The key to these considerations is that context matters: As noted earlier, it is seldom possible to assess highly complex situations – such as social, economic, and political ones – using only singular conceptualisations; rather, it is important to understand existing power and institutional dynamics and tailor interventions to them. Thus, more recent literature has particularly highlighted that specific policy instruments – that is, ways to try to steer policy – may not be directly transferrable or applicable if the variation from the context for which they were originally conceived is too great (and this is assuming it worked well there) (e.g. Sørensen and Torfin 2014; Van Bommel 2014; Jordan et al. 2013). Neither is it necessarily possible to address highly complex situations using only certain specific instruments, or only some of them instead of, for instance, instrument mixes (Lenschow 2014).

Despite this, exactly this approach has been attempted in many contexts: Voluntary market-based means have been assumed to be main means, whereas research increasingly shows that voluntary means on their own are not enough – because they are just that, voluntary (Alberini and Segerson 2002). As we already know from the discussion of knowledge oftentimes not resulting in learning, information or knowledge may be taken up primarily by those who are already receptive to the specific knowledge, stand to gain from it, or the like – it is not generally taken up, particularly if the intended recipients perceive no direct relation to it (e.g. Alberini and Segerson 2002). The situation of voluntary instruments is similar: It may be that only those who are receptive to the features highlighted by the specific instrument may take it up, and these groups may be limited (for instance, those who choose specific certified labels for food or the like). The limitations of voluntary or information-based instruments may also, as some suggest, have been added to by a proliferation of labels, whereby one would in fact have to be highly proficient and knowledgeable (and thereby already interested and receptive) in order to be able to choose – not to mention the difficulties involved with clearly assessing what label or certification, for instance, is green enough, and according to whose opinion (e.g. Bowen 2014). Thus, not only has new market-based governance contributed to the broad discussion on greenwashing; multiple studies have also evidenced the complexity of assessing what processes "really" contribute to environmental impact (on what time scale? to what goals/what minimum level?) (e.g. Marara et al. 2011; Keskitalo and Liljenfeldt 2014).

In many cases, the contextual and institutional dependence of different instruments – and thereby the extent to which they can serve as "solutions" in different cases – has thereby not been sufficiently considered. That is, even if it is known that assumptions characterising the original decision-making and implementation context for which specific instruments and their implementation were defined may not necessarily reflect those for which it might be extended (Sørensen and Torfin 2014).

Recognise social context

As a result, the focus on broad governance instruments – public-private partnerships, participation, or instruments reflecting, for instance, a specific regulative style such as market-based certification or standardisation and the like – may often be based on ideas of what the social context should be and how decisions should be taken, rather than on what the social context in fact is and how decisions are actually taken, as well as limitations in these aspects (such as possibilities to practically implement meaningful participation to make a difference) (Lubatkin et al. 2005; Marara et al. 2011; Stringer et al. 2006).

However, while change is thus not easy to effect (as noted earlier in the chapter) and "solutions" are not catch-all, this does not mean a better understanding of both what theory can say generalisable things about (institutional mechanisms, for instance) and of context dependence (what works where, for instance through understanding both a case and what type of steering it constitutes). It means that the world cannot simply be described in general as "complex" (Olsson et al. 2015, to be further discussed in Chapter 7), but rather that there is method in the complexity. Certain things work in certain states or cases, but not in others. In some states the rule of law matters more than in others, and legal and policy systems may be steered towards specific interests and ways of acting; it is important to understand these types of variations (e.g. Hotho and Saka-Helmhout 2017).

What at the outset seems to be the same type of instrument may also be made workable through differences in implementation in different cases. For instance, a country with a high focus on informational measures (often then backed up with the threat of regulation) may even need to act through mainly information-based or voluntary instruments – however, in such a case, these instruments may work as they are in fact forced (Keskitalo and Liljenfeldt 2014). In countries with a high degree of private steering, where the state in fact is more pulled back or plays a more limited role (Levy 2015), private sector or voluntary instruments may also be highlighted – but perhaps because they are in fact those that "work" and are possible in that system for private actors, absent, for instance, real political abilities to enforce state steering in aspects like the environment.

This does not mean, however, that the same private sector instruments, in an abstracted form (such as standardisation or the like), may work as well in a public sector system like a municipality or region. An example here is how instruments developed for the private sector, such as standards implementation processes and integrated management, typically may not work in the public sector (Klijn et al. 2000; Jørgensen 1999; Keskitalo and Andersson 2017). This is because, among other reasons, the public sector cannot focus its efforts on targeting a single activity (similar to "maximising profits" or effectivising to a single aim in private sector companies) but instead must cover numerous citizen services. Given these multiple aims and the limitations in possibilities for thorough assessment based on new knowledge, public sector processes such as at the local council or municipality level may thus seldom be able to manage "continuous improvement" in multiple areas with limited resources simultaneously, in the same way that a private company with far more delimited aims could potentially do (Burritt and Welch 1997; Talbot 2000; Keskitalo and Andersson 2017).

As a result, approaches typically advocated in environmental management systems and standardisation, including developing "constant improvement" (e.g. Sánchez 2015), cannot be assumed to be applied in all cases, and where they are, it may not be in the way an external observer might assume: Instead, improvements will most likely be in line with already accepted agreements on organisational aims and goals, therefore potentially resulting in incrementalism. In some cases these may be the most workable options, but it is then relevant to be aware of the limitations of the approaches. In a similar way, assumptions concerning how to develop assessments, e.g. in standards aiming towards sustainability integration at the organisation level or the like, may also be entirely too positive in terms of assumed resources and assumed abilities to undertake full-scale assessments based on analysing processes. This is because, for instance, in regular application decision-making parameters are typically not clearly distinguished or identified, but rather operate through assumption and perhaps "muddling through" in relation to organisationally available and enabled information and practice (March and Olsen 1996; Lafferty and Cohen 2001; Keskitalo and Liljenfeldt 2012). Undertaking and following up large-scale standards assessments may thus be too demanding for some organisations (e.g. Le Manach et al. 2020).

Consequently, many of these difficulties involved with gaining traction and change through new approaches, as was the hope with new environmental policy instruments, can be seen as related to problems concerning both a disjointed and incorrectly assumed regulative context and limited comparative assessment to clarify the regulative differences between contexts (Jordan et al. 2013; Lenschow 2014; Marsden et al. 2014; Wellstead et al. 2013). These concerns can be seen as both typifying the problem of going beyond incrementalism in institutional change and as an issue that has been worsened by the limited attention to this contextual variation and institutional aspects in, among other things, large-scale mainstream environmental and climate change research (e.g. Sovacool et al. 2015). Getting at "solutions" requires understanding the institutional context, but understanding the institutional context is no guarantee that "solutions" will catch all: This is an unrealistic assumption regarding what is fundamentally research on human-made systems, whereby humans change in their responses to modifications not least in relation to their interests but also overlap, avoid changing, and build in loopholes for short-term gain or may even not see reasons to change.

Understanding change

This chapter has shown that taking a social science perspective that includes a perspective on understanding institutions can provide a means for understanding change beyond a focus on barriers or opportunities (and beyond a focus on knowledge and learning). An understanding of institutional features supports a better understanding of the limitations to change and also of the various paths to change – along with the fact that there are no "one-size-fits-all" rationalistic solutions that in any magic way would remove the actual complexity and "messiness" of the social world. The fact is that it is research on the social world that constitutes the "hard science": Institutionalised practices, norms, and assumptions concerning the world are hard to change, even after they have been identified – and to make matters even more difficult, they are often seen as a given by those operating within them, which means that such people may not even see a need to change. To this end, one might even argue that it is not the "wicked problem" nature of climate change and other problems of similar high complexity that is the problem; rather, the problem is the fact that we have not always used all the tools (such as varied theories) at our disposal to untangle "wicked" problems. What institutions and organisations are implied by the problem, such as climate change, at the level with which we are concerned (as researchers or as decision-makers)? What do we know about these institutions or organisations and the mechanisms with which they are governed? For instance: What groups of actors and interests are implied and influenced by changes in management that are implied by the problem, and what are their motivations and interests? How are they supported, and how might this support basis be influenced by various changes in relation to somehow managing the problem? And what theories might be applicable – even on a background level – for understanding potential limitations to change? And finally, given this complexity and analysis, what types of instruments could potentially be applied to incentivise change within or in relation to identified institutional logics? Where may the problems lie, and what groups may not be possible to target using specific approaches?

Thus, applying what has been seen here as a broadly institutional understanding, noting the limits of any rationalistic assumptions about policy-making as well as the role of context and the way steering has been developed and "what works" in this context, can provide a way of starting to understand where certain "solutions" in terms of policy instruments may work – and where they may not. Change, even that which is to be affected by *new governance instruments implemented specifically for change*, cannot be assumed to quite simply "work" or "take place". This is contrary to, for instance, climate change literature that has highlighted and often assumed that it will be possible to "mainstream" or integrate consideration of climate change issues in policy (e.g. Klein 2011, compare with Keskitalo and Pettersson 2016; Runhaar et al. 2014). For new governance mechanisms of a more voluntary or procedural nature, a systematic research basis to assess the extent to which, and where, they might "work" (Jordan et al. 2013; Lenschow 2014) would require not only a review of multiple and often varying case studies of different aspects but potentially also original strictly comparative research (e.g. Runhaar et al. 2018). Jordan et al. (2013) note that one reason for the absence of this type of work has been the limited attention both to instrument and to context and assumption; i.e., making explicit the embedded characteristics of the often assumed "neutral" instrument that may make it more suitable in certain contexts than others and investigating implementation in contexts with varying characteristics related to them (cf. Howlett 2014).

Here, it is possible to draw on an understanding of the necessary limitations involved with both mainstreaming and enacting change "just like that" (as assumed in the linear model of scientific knowledge) that are made apparent in institutionalist perspectives. Taking seriously the challenge of understanding and analysing change and decision paths, and acknowledging the limitations to natural science-driven assumptions, could help us move from external and largely descriptive summaries of considerations to describing climate change in the context of social studies at large and within the broader field of the social sciences. There will also exist multiple other theories – and more detailed ones, focusing on specific parts of, for instance, decision-making processes – that are possible to utilise.

To this end, the following chapter will provide examples, from specific cases, of the broad implications of seeing change through a lens focused on understanding the institutions that would be implicated by managing climate change. It will provide examples of the types of results that different theoretical orientations that are not incommensurate with seeing the world as made up of institutions may bring – for instance with regard to understanding how an issue gets on the agenda or how specific ways of governing that differ from assumptions in given countries or contexts can be identified.

Key points

- In social science, theory can be seen as an overarching analytical framework to which the specific results in a case study can be generalised.
- There are possibilities to generalise results, provided one understands the influence of context and process. However, there are also extensive limitations as to what level of "results" as described in theory are generalisable: One has to identify both mechanisms and how they are impacted by the case. One also has to be aware of the existence of many different types of theory, not all of which are intended for real-life empirical analysis.
- Understanding institutional aspects of the social world and how they have developed, as well as are institutionalised, in the present provides us with a better understanding of change and the delimitations to change, as well as the fact that change or mainstreaming is not automatic.
- Thinking clearly about contexts of origin and implementation of different policies or instruments can provide a better understanding of actual possibilities for the choice and implementation of policy instruments in relation to climate change adaptation or mitigation.

Study questions

- What is the purpose of placing the focus on institutions – what does it supply to an understanding of climate change that mainstream framings do not already supply?
- What is an institution?
- What is the role of theory in social sciences as it has been described in this chapter?
- Discuss some of the factors that influence how well different mechanisms – covered in theory or not – are applicable in different contexts.

Notes

1 As noted by Keskitalo and Preston, "the IPCC is perhaps at times dismissive of the value of social theory for understanding adaptation, focusing still on more observational and descriptive insights. For instance, the AR5 report notes that '[a]pplication of social theories may not explain specific cases of human behavior and community decision-making, especially because of the singular importance of the roles of leaders, elites and ideology' (Klein et al. 2014: 920)" (Keskitalo and Preston 2019a: 487). For instance, suggestions in IPCC chapters include that "propositions for change tend to be driven by theory rather than empirically substantiated and tested" (Mimura et al. 2014: 888). On the other hand, it is simultaneously suggested that "adaptation literature would benefit by embracing lessons and experiences of mechanisms for enabling institutional change gained in other policy sectors and past policy interventions" (Mimura et al.

2014: 888); as noted by Keskitalo and Preston, this is "something that would include lessons from what we have here discussed in terms of established social science theorization" (Keskitalo and Preston 2019a: 490, footnote 4). Keskitalo and Preston further note that "while it is true that the actions of actors in a specific case may well result in changes in it progressing differently than another case, this is no reason to undervalue the role of social theorization to in a complex way conceive of the system in and conditions under which leaders or elites may act. In addition, an understanding of historical circumstances cannot necessarily be applied to understand the widely varying situations in the present, even if such an understanding – for instance, in historical institutionalism or path-dependency approaches – can be informative as to what institutions or otherwise conceived system features seem to exert influence, even at different points in time. 'Lessons learned', also, can never be atheoretical: they depend on assumptions as to what features are important and how to define and identify them" (Keskitalo and Preston 2019a: 487).

2 As will be discussed in this chapter, this means that aspects treated in highly varying discussions are related to. The agency vs structure debate largely related to these types of issues (e.g. Giddens 1984), and the institutional perspective taken here, quite simply argue that structure matters (but not that it is the only thing that matters).

3 For instance, a very broad sociological understanding of interaction can entail certain groups being seen as consisting of "relations [that] are historically and culturally specific, dynamic, and polyvalent – that is, charged with meaning through a variety of social ties, including political, economic, and emotional relations. These ties both enable and constrain action" (Ross 2011: 63, as an example).

4 This would imply that institutions bridge actor and structure (see Manicas 2006 for a discussion).

5 This does not necessarily mean, however, that, for instance, increasing globalisation results in a greater variety of rules: Globalisation may instead act as a force homogenising interactions in relation to requirements that allow for relevant power arrangements to be able to interact (e.g. Tsarouhas and Ladi 2013).

6 For instance, path dependency conceptions have been criticised for, among other things, employing explanatory mechanisms ad hoc without necessarily being able to identify what features are path-dependent pre- rather than post-development (e.g. Gorges 2001). However, for the purposes of this volume regarding understanding institutions in a way that provides an understanding beyond the linear model of scientific knowledge, it is mainly the general direction and overarching logics of the theoretical models employed that are relevant. Any reader wishing to employ a specific framework should go into the specific body of literature and internal theoretical debates on the specific framework.

7 The fact that many so-called new governance instruments focused on the environment have been assumed to "just be able to work" whereas practice shows that they do not, often for contextual reasons, is likely a major reason for their limited success in steering towards increased consideration for the environment – and, more recently, climate change – issues since the 1970s (Lenschow 2014). Thus, while attempts have certainly not been lacking, it has been argued that "it is difficult to find the right institutional setup for producing efficient governance solutions" (Sørensen and Torfin 2014: 244) while at the same time avoiding only "passive compliance with broader 'rationalised myths', [rather] than a purposeful strategic process" (Van Bommel 2014: 1161). There is also a lack of studies that take an explicit comparative focus on new environmental instruments and that take into account this context dependence: "[F]ew . . . assess both the wider political context and overarching approach to governance in specific political systems" (Jordan et al. 2013: 162). On the whole, these illustrate that like knowledge cannot be assumed to "just be" transferred, any one instrument, policy, or other type of intervention cannot be assumed to "just work" without both instrument and process as well as broader context being subject to problematisation and study (e.g. Lodge 2014).

Additional readings

Wellstead, A., Howlett, M. and Rayner, J. (2017). Structural-functionalism redux: adaptation to climate change and the challenge of a science-driven policy agenda. *Critical Policy Studies* 11(4): 391–410.

Munck af Rosenschöld, J., Rozema, J. G. and Frye-Levine, L. A. (2014). Institutional inertia and climate change: A review of the new institutionalist literature. *Wiley Interdisciplinary Reviews: Climate Change* 5(5): 639–648.

4

THE COMPLEX OF ISSUES INFLUENCING ACTION ON CLIMATE CHANGE

Examples from forestry and multilevel cases

Introduction

This chapter serves to exemplify the types of results that could come out of undertaking an institutionally focused analysis such as that highlighted in the previous chapter. How can one do an institutional analysis? Drawing largely on the author's own work in climate change adaptation, the chapter will illustrate institutional dynamics, mainly in a case of natural primary resource governance: Swedish and EU forest and forestry systems. Adaptation to climate change in forest systems includes not only the case of forestry but also those of adaptation to invasive species, biodiversity change, storms and the like, and it furthermore requires integration with climate change strategy areas as well as with these numerous policy areas on national, EU, and other levels. This makes adaptation in "forest" in fact a highly complex area, relevant as an illustration of the complexities of adaptation "mainstreaming" across a historically developed sector that today relates to several different policy fields. It will thereby also serve as an example that is returned to in other chapters, in order to illustrate different theoretical aspects and considerations.

The chapter will also take examples from multilevel adaptation attempts in Sweden and the UK with regard to local, regional, national, and EU levels, and the difficulties of acting in these similarly institutionalised systems. Through this, the chapter will also illustrate the different approaches to adaptation in different countries and constituencies, which may include institutional dynamics at national and lower levels.

The aim of this chapter is thereby mainly illustrative: to highlight what an understanding centred on institutions and institutional development can do to help us understand possibilities for adaptation and mitigation. The idea is that the chapter, through the case of forestry and a discussion of different levels, will be able to present the types of understandings that are also relevant in any other sector,

DOI: 10.4324/9781003043867-4

country, or case in order to really, in an institutionally focused way – making an understanding of the institution central to truly understanding in a contextual and realistic way – conceive of the possibilities for adaptation and mitigation. The chapter will thereby relate to and contextualise many of the issues taken up in the previous chapter, such as context dependence, formal and informal institutions, different governmental logics, incremental or radical openings for change, and the role different policy instruments may play.

In order to extend an understanding of the role different theory can play for understanding different parts of phenomena, the various cases will also be discussed through the lens of theoretical frameworks that are to varying extents related to a focus on institutions, in that they all base their analysis on the real-life, power-influenced, character of interaction. These include theoretical frameworks that specifically highlight, for instance, institutional path dependencies, discourses and governmentality, and certain understandings of multilevel governance.

Understanding the factors that impact adaptation (and mitigation) in forestry: the Swedish case in perspective

The previous chapter highlighted the role of understanding the institutional development and specificity of different sectors, actors, and interests in helping us understand the possibilities for climate change adaptation and mitigation in relation to the chosen area. This means that understanding adaptation and mitigation possibilities is context-specific. Thus, this section will focus on forestry in Sweden but will also to some extent contextualise the narrative in relation to other cases.

The previous chapter also highlighted the importance of understanding the role of the sector, or any other chosen case, and how it has developed historically. This involves, for a natural resource sector, not only understanding its use but also perhaps national and population development, industry, and interests as well as organisational bases. For a resource sector this would include how the resource is used; by whom; through what organisational, legislative, and regulative structures; and to what purposes. Information on these types of factors can be found in a wide range of literature in several disciplines and can also be developed through, for instance, legislative, policy, or interview studies; studies that have applied these and other methods are used as the basis for this illustration (and referenced in the following section).

The significant economic role of forestry in Sweden

A crucial part of institutional analysis involves knowing what uses (and associated actor complexes and interests) institutions were developed for. In Sweden, the economic role of forestry has been pronounced through time, and it is crucial to understand in order to understand the sector's configuration. An interesting fact in the Swedish case is that, throughout history, a major focus in how the forest industry and forestry have been developed has involved economic export income

(Kunnas et al. 2019). This can be seen in that early on, forest wood production was valued for its role in providing charcoal for mining production. When the value of forest shifted into wood production per se in the early 1800s, this was largely the result of export valuation changing: foreign buyers now required not only minerals but also wood (Kunnas et al. 2019). However, while the role of wood production shifted, its purpose did not. Thus, the earlier use of forest for developing minerals sold on an international market and its later use for developing wood sold on an international market were similar in giving an economic purpose to forest production.

This means that while local actors and households in Sweden, as elsewhere, used forest largely for historical local uses such as for firewood, hunting, berry and mushroom picking, and the like, the role of industry developed in parallel in a way that it did not do in all other countries. A comparison can be made to the Mediterranean, for instance, where forest is less plentiful and the forest structure different. Sweden is mostly forest-covered, making forest-based sectors a large basis for both population and industry, and resulting in forest industry production to this day remaining a major focus (Keskitalo et al. 2013). This can be compared with the fact that in "Southern Europe the non-wood products and protective functions, especially the water related services are often dominating" (Parviainen 2006: 70).

Perhaps as a result of this type of development, Swedish forest is thereby largely managed throughout its life, from planting to logging. Swedish forest is marked by relatively intensive semi-natural forestry (McDermott et al. 2010), focused on the production of even-aged stands of single dominant tree species, based largely on planting (Axelsson and Östlund 2001). Today, Sweden is also one of Europe's largest forest countries in terms of production and the role of forestry in GDP and export value, and it is the world's third-largest exporter of sawn goods and pulp and paper, respectively (SFIF 2020). These numbers illustrate this very large role of forest industry even on the national level – and in fact also the continued economic role of forestry.

A historically rural connection institutionalising production thinking among forest owners

At the same time, Sweden (along with Norway and Finland, for instance) is historically a highly rural country, with low population densities spread across the country. This, along with specific historical development lines, has led to the population even to this day retaining a relatively strong linkage to rural areas. This can be seen in, among other things, recreational and ownership patterns, for instance in the large proportion of Swedes (along with Norwegians and Finns) who either own or have access to second homes (Back and Marjavaara 2017; Rye and Gunnerud Berg 2011; Vepsäläinen and Pitkänen 2010; cf. Keskitalo et al. 2017a). What is more, this background has also fundamentally contributed to about half of Sweden's forests today being owned by some 330,000 small-scale or non-industrial family forest owners (Keskitalo et al. 2017b).

Given the large areas and multiple interests, there are conflicts, e.g. between environmental NGOs and forestry as well as forestry and reindeer husbandry and regarding the role of property ownership in the system (e.g. Lazlo Ambjörnsson et al. 2016). However, the fact that forestry is of such prominence legislatively, and still very much operational and not subject to more conflict than it is (although there is conflict), could potentially be attributed to these ownership patterns as well as to the fact that many uses are nevertheless allowed on private land. This includes berry picking, hiking, and the like, which are allowed through the Swedish Right of Public Access. The fact that so much of the forest is owned by small-scale private forest owners also means that a relatively large number of people beyond the industry are to some extent invested in forest, and this also goes for the relatively large hunting and fishing interests. The interlinkage between various private uses can also be large – for instance, individuals who are reindeer herders may also own forest or work in the forest – making conflicts more sectoral than individual (Keskitalo 2008b).

These facts, including that half of the domestic resource is owned by small-scale owners spread over a large area with many interests, while Sweden still maintains an internationally prominent forest industry with heavy use of domestic wood, are thus in themselves remarkable. Something – institutions or institutional configurations that enable this – must thus have developed in order to assure that individual owners would provide the forest industry with wood in the way they do. In literature, it has been noted that these simultaneously strong roles of individual owners and the forest industry could potentially not have been the case without the forest industry being well established in a developed linkage to small-scale owners as well (e.g. Lönnstedt 2014; Keskitalo 2017).

- ***The role of forest owners' associations***

One way in which this has taken place is through forest owners' associations (e.g. Haugen et al. 2016; Lönnstedt 2014). In European comparison, and perhaps also internationally, Sweden and to some extent Norway and Finland stand out as countries with very strong forest owners' associations (Lönnstedt 2014; Keskitalo et al. 2017b). Among other things, the now three large forest owners' associations in Sweden, owned by their members, contact new forest owners with offers to provide them with forest management and logging services – in fact making it possible to be a forest owner without knowing forestry (Andersson et al. 2020).

At present, the forest owners' associations also have outreach and members' meetings in the large cities, to among other things gather new forest owners (who may have inherited forest or otherwise do not know forestry per se). Direct contact with forest owners, for instance in relation to the age of their forest, may also be taken by the large forest industries offering to sell management services similar to those the forest owners' associations offer (Andersson et al. 2020; Andersson and Keskitalo 2019).

- ***The role of available forest, population statistics, and support organisations***

This type of development, schooling new forest owners into the model of production for industry (even in the case of smaller holdings), is supported not least by Sweden's historically well-developed data on both population and forest (Keskitalo 2017). Quite simply, forest areas and forest use are much easier to track in Sweden than in many other countries: Who owns forest, where, how much, what the condition of the forest is, and the like, are catalogued. Historically, this availability of database information, not only on forest but also the population on a detailed individual level, may have been developed in relation to Sweden's large geographical area and low-density, largely rural population (i.e. as an instrument for the state to gauge population, for instance for tax purposes, and continuing into other sectors) (Lidestav et al. 2017).

Thus, compared to many other countries, Sweden has well-developed and accessible data that can serve forest companies and forest owners' associations in providing knowledge about forest owners, further enabling the system of large forestry production even given the dispersed ownership.

In addition, multiple institutions have developed, supporting the system of a well-institutionalised forest industry despite dispersed ownership, for instance the measuring system for enabling the forest industry or associations to calculate the price of timber in a similar way across the country (cf. Stjernström et al. 2017).

- ***The role of "the Swedish forest model"***

Crucially, as perhaps one of the strongest factors, forest-relevant industry and the state also did not develop in isolation from each other. Instead, the institutionalisation of a specific model of acting in forestry has occurred in relation to the state (not least given the relatively large economic role of forest production even on the state level). In international literature, Sweden has even often been seen as a corporatist state: one where a strong state was supported by a strong linkage to organised groups, including economic interests and sectors; it is unlikely that this linkage was not pronounced in earlier periods as well (e.g. Ogilvie and Cerman 1996 eds; Öberg et al. 2011).

Taken together, then, forest industry and forestry in Sweden thus comprise a historically developed institution that cannot be well understood without also understanding these types of historical, structural, and ownership prerequisites. Sweden, thereby, is not a "typical" forest case (if any such might exist) but rather more of an atypical case, marked by a highly specific historical development which has also given the forest industry a prominent place that continues – well institutionalised – to this day (Keskitalo 2017; Lidestav et al. 2017).

Tracing institutional paths

While this case illustration does not necessarily mean that one always has to dig this deep to be able to understand factors that influence what actors, interests, and institutions influence adaptation and mitigation possibilities and paths in the present, it does mean that how a sector or other case has developed in its specific national setting will play a role in how, and in relation to what established interests, it will act in relation to new challenges or demands as well. For instance, the forest sector in, say, Italy will have developed in highly different ways, with very different ownership and even purposes of forest: In this case, it might mean that forest industry is far from as strong or well-institutionalised an actor, and other interests are then perhaps dominating the forest use (Keskitalo et al. 2013). This would fundamentally mean that forest sector adaptation possibilities and paths will also differ between Sweden and Italy (as will be discussed later).

The way the role of forest industry manifests in Sweden today, for the purposes of how these types of situations can be traced in other countries or cases, can be traced in the legal framework and, for instance, in legal, policy, and forest sector literature. It is relevant, however, to be aware of both what the formal requirements in law are, for instance, and how they are implemented in practice (meaning that it is an issue of analysis rather than merely looking at the formal legislation).

- ***Legislative development in context***

We have already noted the significant role of wood production, and Sweden's first formal Forest Act (in 1903) focused on wood production (although laws since the 1200s–1300s had regulated production based on forest, then mainly for mineral purposes; i.e., wood used for powering furnaces for mineral production) (Kunnas et al. 2019). Similarly, we have noted that wood production remains the main purpose of forest acts to this day. This has been the subject of much discussion. In particular, the amendment in 1993 to the 1979 Forest Act preamble was to formally place what is often called "production" and "protection" (or conservation) aims on the same footing. However, in literature, the formal equality in the preamble has largely been assessed to be delimited in its effect, resulting in wood production remaining the main aim of forestry. Not unrelated, the 1993 Forest Act also introduced the principle of "freedom under responsibility", by which forestry is deregulated and expected to achieve certain goals but can also choose how to accomplish these goals (largely within a production framework, Appelstrand 2007).

These legislative settings are important as they indicate the role and strength of forestry in legislation. Not only does "freedom under responsibility" make advice or informational measures one of the principal – albeit limited – means by which forestry can be influenced, but it also means that the role of existing practices will have a major steering effect on the way any policy is implemented and that the role of protection is to be implemented largely through means developed by the sector itself (Keskitalo and Pettersson 2012; Appelstrand 2007). Established practices and informal norms in forestry are thereby in fact attributed great importance and have also been seen as providing the basis for what is often called the "Swedish forestry model" (Appelstrand 2007; Törnqvist 1995), i.e. the social regulatory practices of forestry in this context of historically developed forest industry (Andersson and Keskitalo 2019).

The specific way in which forestry in Sweden is institutionalised and specific ways of governing are in fact recognised can thereby be seen as related to its strong entrenchment among industry, forest owners' associations, and others, and how forest practices are then in fact developed by what these actors do (practices) within relatively well-institutionalised frameworks largely governed by historically developed institutional logics.

Another example that has been noted is that the forestry sector is seen as *ongoing land use* (Swe. *pågående markanvändning*); that is, land use that is seen as foundational – not to be negatively affected by other interests – even if other uses co-exist in the same area (Kunnas et al. 2019). This role is visible in the fact that there is compensation for this land use, for example for restrictions imposed by the public that impede "normal and natural rationalization measures" in forestry (SOU 1971: 75, 107, 108, quoted in Kunnas et al. 2019: 67). Interestingly, this means that measures, e.g., to reduce environmental impact, cannot hamper forestry as an ongoing land use, which may then subordinate protection to production despite the formal focus on both aims in the Forest Act.

As a result, forest protection typically involves instead creating nature reserve or habitat protection areas that require compensation to the landowner (Kunnas et al. 2019, as enforced in recent legislative processes). Again, this highlights the assumed role of production. Implications of the standing of forestry also include that reindeer husbandry, by some assessments practised on 40% of the Sweden's area and much of this on land owned by forest owners but with reindeer husbandry as a user right despite being regulated in its own Act, is largely governed by forestry legislation. This is because, among other things, reindeer husbandry cannot impede forestry as an ongoing land use; thus, consultations (as a type of stakeholder interaction) between reindeer husbandry and forestry, mandated by law but in extended areas also by market-based voluntary forest certification, have been found not to play any greater decisive role over time (Kunnas et al. 2019; Axelsson et al. 2019; Keskitalo 2008a).

Other legislation besides that described in this text can also be traced to illustrate the ways in which the sector's role manifests legislatively. As one example, forestry and mining, the two historically important export sectors, are today the two large-scale land uses in Sweden that fall under sectoral legislation that stands apart from the Environmental Code; as special legislation supersedes general legislation, the two sectors are thereby not as integrated in a broader environmental context as they might otherwise have been (Kunnas et al. 2019). Similarly, spatial planning, often seen as the local monopoly of municipalities in Sweden's strongly decentralised system, is generally not integrated with forest planning, despite large areas of municipalities in many cases being forest (Stjernström et al. 2017).

In legislation, forestry is thus well encased not only as a sector with a practical possibility to focus on production under "freedom under responsibility", but as an ongoing land use under specific legislation that is seen as primary to other land use. This type of formal institutionalisation of forestry must be considered along with the practices that together make up the "Swedish forestry model", such as the forest owners' associations and large individual and family ownership of forestry that thereby support and also legitimise the model.

- ***Voluntary instruments and what they imply***

Forestry can thereby be seen as made up of both formal and informal institutional mechanisms that work in relation to specific historically developed logics. We might thus not expect that voluntary or other means that are applied will be at a strong disjunction with these (or else would likely not be applied in the sector).

In the previous chapter, we saw that voluntary instruments could be assumed to be applied when they were incentivised through broader structures, and then in the way these broader structures provide incentive. The use of voluntary market-based instruments in Swedish forestry illustrates just that, and it can be seen as a means by which production values are supported by means other than legislation.

In production forest areas (the term itself illustrating the assumed production role of forest in wood production), environmental values are typically largely managed through voluntary market-based certification (FSC or PEFC); that is, voluntary governance systems. Through these, for instance, a percentage of trees including broadleaf and dead wood with environmental value are retained at logging (e.g. Keskitalo and Pettersson 2012). Forest owners are compensated for not taking out all the forest by getting a slightly higher price for their wood, and this system has become so accepted today that, despite formally being voluntary, it almost constitutes a market requirement for the industry, who then also has the incentive to pass the system on to forest owners (Stjernström et al. 2017).

Forestry can thereby be seen as – and has indeed internationally been seen as – a prime example of how market-based voluntary instruments can work (e.g. Cashore et al. 2004). However, as this case illustrates, they may work under highly specific conditions. Internationally, certification became the way for the timber industry to illustrate its sustainability under conditions whereby work towards an international forest convention had stranded. This was a factor that perhaps influenced the possibility that certification, even given the emergence of different forest certification systems, could become a requirement on certain markets. In Sweden, then, as a major exporter, it should thus not be surprising that certification plays a significant role. However, while the forest industry sees certification as a large step forward, certification is regularly criticised by environmental NGOs for not going far enough (the change being mainly incremental, as one might expect from a market-based voluntary system) (Cashore et al. 2004; Keskitalo and Liljenfeldt 2014), and criticism has also been raised regarding the limited role of consultations both legislatively and as extended under certification by reindeer husbandry (e.g. Keskitalo 2008a).

The role of forestry, in relation to environmental but also other issues, is thereby under continuous discussion; however, its role particularly in relation to production – but also other sectors – must be seen as well encased both legislatively and in relation to interpretations in private governance (and how far this can go).

Again, both legislation and the application of different instruments, together with an understanding of the practices and informal institutions that support them, can offer an opportunity to trace the role of forestry in relation to other sectors.

Consequences of the way the institution of Swedish forestry is set up for adaptation and mitigation

Given not just the economic role but also the long-term characteristics of a sector like forestry, an external "rational" observer might then have thought that forest would be one of the first sectors in Sweden (and elsewhere) where adaptation to climate change would be implemented. After all, in Sweden, forest stands for perhaps 70–90 years before final felling – far into a climate-changed future. And due to both the great economic importance of forestry in Sweden and the fact that almost all forest today is planted and managed – making it possible, for instance, to choose a tree species and then manage it throughout its life – one might have assumed that ensuring that forest thrives and stands throughout changing climate conditions and extremes would be paramount (Andersson and Keskitalo 2018).

However, this might be an assumption one would make before taking into account the potential economic costs of events that may seem uncertain and that to some extent challenge management practice in forest industry. Climate change impacts of relevance for forestry in Sweden include the benefits of a longer growing season (Andersson et al. 2015) as well as the risks of damage resulting from storm events, drought, and other water stress and insect or pest outbreaks (Commission on Climate and Vulnerability 2007).[1]

Some adaptations in relation to these impacts may be relatively easy to undertake. Forest, being largely planted, benefits from plant material being to some extent selected for future climate stresses. In addition, a longer growing season may make it possible to log earlier (shortening rotation times). Adaptation to storm risk could also include, for instance, limiting the sharp edges of forest plantations (Andersson et al. 2018; Andersson and Keskitalo 2018; Keskitalo et al. 2016). However, some of the most important adaptations that might be undertaken would include developing more mixed forest, as this would be more resilient to storms, drought, and pest/insect outbreaks than the present monocultures centred on specific species. Improved management for storm risk might also include a greater focus on continuous-cover forestry and shifts in the species planted. These types of adaptations would mean switching from a focus on business as usual and planting of the tree species that are most economically valuable, which are largely logged through clear cut, to a focus on resilience under stress whereby mixed forest would include not only the most economically valuable species but also potential changes in logging practice (Andersson et al. 2018; Andersson and Keskitalo 2018; Keskitalo et al. 2016). More far-reaching adaptations would thereby pose a cost to the sector in relation to business as usual.

In addition, there are only limited broader (legal or national policy) requirements for adaptation. At present (2019–2020), Swedish adaptation policies with regard to forest, potentially in relation to a focus on freedom under responsibility in Swedish forestry, do not express strong binding requirements, even under the new climate policy framework and government adaptation strategy (Government Offices of Sweden 2017; Government Offices of Sweden 2018). In general, the Swedish Forest Agency is required to provide information on adaptation and has run information campaigns and provided advice, for instance (Commission on Climate and Vulnerability 2007; Government Offices of Sweden 2009; Keskitalo 2011). Following the development of the new Climate Policy Framework that came into effect in 2018 and the government adaptation strategy developed at the same time, the Swedish Forest Agency is in line with other state authorities required to develop action plans and report on progress on climate change adaptation in forestry, among other things (Government Offices of Sweden 2017; Government Offices of Sweden 2018). This means that decision-making on adaptation remains largely up to the private, non-industrial forest owners who own half of the Swedish forest land (and who may not be well informed about adaptation) as well as to industry.

Thus, with regard to which adaptations are mainly being considered, it has been identified that these could almost be foreseen based on the institutional context of Swedish forestry (Kunnas et al. 2019). Potential climate change adaptations have been described as being in line with Swedish production models, focused for instance on planting quick-growing exotes (such as pinus contorta, for which plantations already exist) and increasing rotation (logging earlier), which would benefit industry (e.g. Keskitalo 2011; Keskitalo et al. 2016). Less focus is placed on mixed forest, continuous-cover forestry, or shifting planting from high economic value species that would limit the maximum economic gain under business-as-usual conditions (Andersson et al. 2018; Andersson and Keskitalo 2018; Kunnas et al. 2019). In a study interviewing representatives with a formal role in adaptation from all major forest industry actors in Sweden, it was observed that interviewees generally noted that little had been done on adaptation per se and that adaptation-relevant actions undertaken generally involved aligning with existing environmental requirements, for instance requirements related to forest certification systems (for instance retaining a proportion of broadleaf) and to the Water Framework Directive (such as increasing consideration in relation to water) (Andersson and Keskitalo 2018). Adaptation-relevant actions that had been undertaken, for instance in relation to the large storms that had taken place in Sweden (storms Gudrun and Per, in 2005 and 2007, respectively), were also seen as related more to these events than to forward planning in relation to climate change. As an explanation for how they acted in relation to adaptation, interviewees continuously emphasised the role of production requirements in the system (Andersson and Keskitalo 2018). As a result, it has been noted that:

> [a]long these lines, climate change adaptation may be currently regarded as representing more of a coping type of development, modifying the system along the line of existing orientation, rather than involving more far-ranging

> adaptations that would serve to change Swedish management systems logic or implement entirely new adaptations.
>
> *(Keskitalo et al. 2016, section 4, para. 2)*

And it was not only adaptation that was conceived of in this type of economic context but also mitigation. Interviewees in the 2019 study saw the production focus as highlighted in relation to mitigation as well, among other things as an "enormous opportunity for business" (in the words of one interviewee, Andersson and Keskitalo 2018). It has also been noted that mitigation has been emphasised more than adaptation in the Swedish context, largely as the forest can be seen as a carbon sink and thereby helps Sweden stay within its carbon budget, or as an important material with regard to substitution (which would provide an additional potential high-value output area for forestry, at a time when paper demand for press has been decreasing) (e.g. Lundmark et al. 2014; cf. Kurz et al. 1997, 2008; Klapwijk et al. 2018).

Thus, it has been noted that a "focus on the intensified production of forest biomass . . . to enhance climate change mitigation . . . [is an] expectation . . . underpinned by the fact that the long-term net annual growth and product-use strategies in managed forest systems are integral to climate change mitigation" (Klapwijk et al. 2018: 241; cf. Andersson et al. 2018). These areas would support export and production aims as well as the role of the Swedish forest industry, and support forestry business-as-usual (Kunnas et al. 2019).

The Swedish case could thus be seen as illustrating that the historically developed structures existing in the present – and embedded legislatively as well as in practice – may be the best predictors of adaptation and mitigation behaviour absent strong, direct, and clear, for instance economic, incentives to the contrary. The Swedish forestry structure – with its strong focus on production – could be seen as "coming close to presaging what mitigation options will be highlighted" (Kunnas et al. 2019: 72, potentially also relevant to adaptation).

The case study description and analysis can also be seen as illustrating the prominence of specific instruments before others: While the Swedish case is presently marked by formal deregulation in forest policy and a prominence of certification, the well-institutionalised legislative requirements regarding forestry promote largely the same aims as in production forestry – a prominence of forestry land use and functionally the continued economic role of forestry, as a framework within which forestry itself is to develop consideration or implementation of more far-reaching environmental or other aims. Seeming deregulation, therefore, is in fact complemented by a strong legislative framework towards continued forestry production, and the use of voluntary instruments that are crucial in, for instance, supporting the economic market structure of forestry with high export.

Thus, while the Swedish case may be particular, it illustrates the types of institutional mechanisms whereby historically developed institutional structures also govern adaptation and mitigation development and the orientation it takes. In other countries as well, the way approaches to adaptation or mitigation are in fact taken in relation to existing logics and use aims can be identified. In Italy, where the focus in forest is less on industry than on local production of multiple uses, one approach that has been mentioned is "passive adaptation"; that is, leaving the forest to adapt on its own (Keskitalo et al. 2013). This approach can be seen in relation to a logic whereby the forest and the production in it are not nationally important or targeted per se, and thereby forest values can be left to whatever consequences occur under climate change; that is, focused on coping more than adaptation. Across EU countries, adaptation to climate change in forest may follow different paths, for instance related to adaptation strategies being developed on local, national/federal levels, and to strategies being developed as general or specific to forest (Keskitalo 2011) as well as to the national and regional historical context of forest development and use.

Applying more specific theoretical lenses to the Swedish case

While this analysis focused on very broad institutional factors and how forestry can be seen as institutionalised, applying other more specific frameworks (or at least illustrating what such applications could highlight) can demonstrate the ways various features can be highlighted through the use of different theoretical orientations.

In general, forestry can be seen as exhibiting traits that could be interpreted in relation to multiple theoretical frameworks, and this will be briefly illustrated below. In general, researchers typically choose what theoretical framework they want to apply in relation to the mechanisms they are most interested in investigating. However, using different theoretical frameworks can also provide a broader understanding of an issue – particularly if one clarifies what specific features different theoretical orientations are applied to highlight.

Historical institutionalism and path dependency

In broad relation to a focus on discussions relevant to historical institutionalism or *path dependency*, the forestry system generally exhibits a focus on production use over time, and the earlier narrative illustrates how this is both maintained and encased legislatively, as well as given room and acknowledgement through foci on "freedom under responsibility" and the "Swedish forestry model". The narrative also illustrates the role of multiple actors or organisations whose interests converge around, or who at least gain benefit from, the production aim – the state, industry, forest owners' associations that sell services and provide timber to industry, and forest owners who are schooled into the forestry model: "a network of interconnected forestry-related organizations, instruments and practices" (Kunnas et al. 2019: 73).

This would mean that there are interconnected, potentially self-reinforcing systems and subsystems, which would raise the costs of change for actors within it (Pierson 2000). As noted previously, it could even be considered whether the present large focus on "freedom under responsibility" is in fact made possible by the well-institutionalised structure of forest industry: The state could be said to know what it's getting in terms of production, even under deregulation, but can to a lesser extent be blamed for the forest industry not living up to, for instance, increasing environmental requirements as they may be phrased by other parties (such as environmental NGOs).

The well-institutionalised production model can also be seen as selecting what other options, for instance with regard to adaptation or mitigation, become the most relevant.

A focus on specific paths in the face of climate change could potentially be seen as lock-ins of the production model, selecting options that are beneficial for the production aim while leaving out those that are not, in line with a business-as-usual rather than resilience-under-stress model (Andersson and Keskitalo 2018). It has thereby been considered that, "[a]s domestic structures for forestry are so well institutionalized, their standing can thus, presumably, primarily be shifted by external events – such as the current changes in the demand for raw materials and forest products by international markets, or by factors related to climate change" (Kunnas et al. 2019: 73).

Foucauldian governmentality

These same traits could also be seen through a Foucauldian lens, as briefly mentioned in the previous chapter. While seldom seen as an "institutionalist" approach, a Foucauldian approach perhaps has pattern similarities to a focus on institutions, as it highlights the role of system or structural (institutional) characteristics in delimiting the room for manoeuvre. Foucault developed the concept of *governmentality* to describe the "conduct of conduct", or governing mentalities, assumptions, or rationalities that "shape, guide, or affect the conduct of some person or persons" by making them seem logical (Gordon 1991: 2). This concept could be applied as one way to understand the means by which the interconnected network of forestry-related organisations work to, among other things, continuously provide timber to industry and school new forest owners into production. In line with Foucault's more general orientation towards discourse, governmentalities are seen as structuring what is seen as knowledge: "Each society has its regime of truth, its 'general politics' of truth: that is, the type of discourse which it accepts and makes function as true" (Foucault 1991: 131, quoted in Winkel 2012: 82). These mentalities are then implemented through the application of, for instance, specific "technologies" (ways of steering action) that fall within the logic of the mentalities (e.g. Rose 1996; Rose and Miller 1992).

Contrary to a focus on knowledge as a given truth (the way it is assumed in the linear model of scientific knowledge), then, this type of conception would highlight that there are specific assumptions as to what is true, as structured by the institutions through which these "truths" are developed. For the Swedish forestry system, this could include a focus on production, monoculture, and industry as an aim, implemented for instance through technologies of expertise and service provision (Andersson and Keskitalo 2018). Not least, such a conception might also highlight the role of forest certification as a technology – an instrument whereby forest industry practices can be continued and supported, even if under concessions in relation to shifting requirements on the sector (such as, here, the market as a fundamental impact on economic viability that requires that specific environmental measures be taken to allow for the certification) (Andersson and Keskitalo 2018).

This type of approach would add to one focused on, e.g., path dependencies by focusing on the difficulties that thereby exist even in providing contrary information to knowledge within the system: Focused on providing wood to forestry, the industry system including forest owners' associations might, for instance, not provide clear adaptation or conservation advice to its individual forest owner members, but instead highlight the mechanisms or technologies that are made more relevant through these systems, for instance certification (Andersson and Keskitalo 2018). In addition, it might be difficult to reach out with information that implies a need for large-scale change in the system (such as mixed forest with less benefit to currently economically valuable species) (Andersson and Keskitalo 2018).

Multilevel governance

The Swedish forestry case could also be seen as one of *multilevel governance*. The concept of multilevel governance is often seen as positing that various actors, such as international, EU, and national legislation and policy, as well as private and NGO initiatives, may play a role in governing and steering various issues (e.g. Marks and Hooghe 2004). This type of theoretical framework was largely developed to conceive of the notion that it is no longer only the state that in practice steers development, but that other actors may also functionally play significant roles (Marks and Hooghe 2004). Multilevel governance conceptions have often also been drawn upon to highlight the different democratic implications of this shift, arguing for the need to increase direct democracy, for instance through participation (Hysing 2009).

However, given what we have already learned about institutions, for instance in the preceding chapter and the illustration in this chapter, based on a focus on institutional aspects we might not assume that individual participation would necessarily result in a change in embedded structures with much stronger and well-organised actors, among the full variety of actors that may exist within multilevel governance. Perhaps not surprisingly, then, in writing about multilevel governance in relation to forest in Sweden, for instance Hysing (2009) illustrates the strong role of the government in private governance in forestry in Sweden: that the state may in practice delegate the follow-up of rationalised forest policy to industry, by seeing certification as a means to implement delegated forest policy aims (Hysing 2009; Johansson and Keskitalo 2014). Thus, one could rather conceive of multilevel governance from a more institutionally focused perspective, as "constitut[ing] a system of steering that needs to be understood in terms of how it plays out in its entirety in relation to different questions and issue areas" (Keskitalo and Pettersson 2016: 56; cf. Marks and Hooghe 2004; Pierre and Peters 2005).

Conceived of in a multilevel framework, however, the limited focus on adaptation in forestry would not be seen as unconnected to the supranational and national policy environment on adaptation at the time; such a focus might then necessitate reviewing the Swedish forestry sector development on adaptation to an even greater extent on a Swedish and an EU level. On the EU level, it can be noted that a relatively non-formalised approach to adaptation (Ellison 2010) was present until the 2013 EU strategy on adaptation to climate change, which is aimed at European states adopting comprehensive adaptation strategies, supporting better informed decision-making, and promoting adaptation in key vulnerable sectors such as forestry (COM (2013) 0216 final; cf. SWD (2013) 131 final, SWD (2013) 132 final). However, given its focus on general guidelines, the strategy largely provided general non-binding aims rather than legal requirements. This is because the supranational level in the case of the EU does not at present mandate action in forestry (for reasons not least involving negotiated state sovereignty and what areas should fall under EU and national prerogative or formal competence) (Keskitalo and Pettersson 2016). (The strategy is presently under revision, with a new version to be launched in 2021.)

On the Swedish level, as noted, climate change and more specifically adaptation have been managed mainly through the Commission on Climate and Vulnerability (2007) and the Climate Bill (Government Offices of Sweden 2009), as well as most recently the Climate Policy Framework (Government Offices of Sweden 2017) and the national strategy for adaptation (Government Offices of Sweden 2018). The 2007 Commission – like much adaptation work in other countries at the time – largely focused on water risks (Keskitalo 2010, ed). Developed largely as a result of flood risk concerns in southwestern Sweden, the Commission discussed forestry mainly as one among many sectors in a large review. It also discussed measures in forestry in relation to how it treated multiple other areas, to be subject to sectoral coordination by its sectoral agency (for forestry, the Swedish Forest Agency) (Keskitalo 2010a).

In effect, however, as noted earlier, this left adaptation up to decisions among the individual forest owners and industry, in line with the "freedom under responsibility" approach (e.g. Keskitalo et al. 2011). Similarly, the 2009 Climate Bill also treated adaptation in forestry as a matter for the sector rather than the state (in line with the general deregulation approach in forestry regulation, described earlier). In 2018 under the new Climate Policy Framework and adaptation strategy, to some extent, the general "mainstreaming" approach in relation to already established aims for the forest sector continued, and so far it does not implement specific strong requirements given the strong deregulation in the forestry sector. As a result, with neither the EU nor national level so far having provided strong steering measures, it is thus perhaps not a surprise that the more limited informational measures on adaptation have had limited impact and that the strongest impacts (on coping rather than adaptation) have been from those requirements that have been stronger or more influential, such as forest certification or the Water Framework Directive as already discussed (Keskitalo and Pettersson 2012).[2]

Present institutional frameworks and dynamics, highlighted in the use even of varying theoretical frameworks and the role of larger-scale systems, can thus be seen as an illustration of why relatively little has happened in regard to adaptation: So far, adaptation has been governed largely through voluntary or informal means that have added on to existing structures – on different levels – that steer far more strongly towards other aims. This analysis could be seen as underlining the approach that to understand adaptation and mitigation decision-making, it is more important to understand the sector-, country-, and system-embedded (even supranational) logics than to focus solely on climate change impacts or future scenarios of potential development (cf. Beck 2011).

The variation in approaches to climate change adaptation on local, regional, and national levels

The role of specific institutional logics on a national and, for instance, sectoral and subnational scale that was discussed for the Swedish forest case can also be seen in studies on climate change adaptation in comparative national contexts.

Previous studies on general climate change adaptation in the EU have illustrated that, due to different institutional setups and assumptions, different countries go about climate change adaptation very differently and, even in cases in which similar programmes are set up, implement them differently and to different extents (Keskitalo 2010, ed). So far, we have reviewed one highly specific case in the forestry sector. However, studies on climate change adaptation, especially early ones, have largely involved the fields of water and particularly flood risk rather than for, instance, forestry or areal land uses per se (cf. Keskitalo 2010, ed). Among other things, this seems to have been the result of the significant focus particularly on increased flood risk at the time, with floods also having clear effects among constituencies fearing for their homes (Keskitalo 2010, ed). Nevertheless, this confluence of flood and climate concerns and how it came to be emphasised was also not automatic. Studies show not least the role of factors typically emphasised in agenda-setting literature – such as events like a flood event, development of policy, and supporting politics – and the need for all these factors to be present for adaptation policy to be established (cf. Kingdon 1995). However, they also illustrate the highly varying role of the development of formal policy, sometimes with more formal or layering functions ("Now we've done something about this") than practical ones (guiding decision-making in very costly and difficult-to-coordinate fields) (Keskitalo et al. 2012a, 2012b).

Taking examples from a comparative multinational and multilevel study of adaptation policy development in Finland, Italy, and the UK as well as Sweden (published in 2010 and after), it can be seen that countries have worked in different ways with adaptation strategy development. Like in the analysis described here, the development was largely steered by existing structures (here, for instance boards with overarching decision prerogatives focused less on climate than on potential cost) and practices in the country (Keskitalo 2010, ed; Keskitalo et al. 2012a, 2012b). This section thus provides a comparative discussion of different national contexts and different levels, and it applies different frameworks to the study of all four countries. This is intended to highlight the role of national organisational systems as well as that of different levels and historical contexts in how responses to climate change may be formed in different countries. It can thereby serve as a further illustration of, for instance, issues of context dependence, discussed in the previous chapter.

In *Sweden*, thus, adaptation has progressed through the development of a governmental investigation into a recent formal framework and strategy (2018, discussed previously and built largely on the UK example, too recent to assess in detail here). So far, a major concern in the national context has been that climate change adaptation responsibilities between municipalities and the state have been delimited in relation to the existing structure in Swedish decision-making at large. The development of climate change adaptation was the strongest early on concerning water and flood issues, which had also been driving factors in the inception of adaptation work; this was largely as a result of a regional-level appeal to the state citing issues of national security, thereby gaining a higher position on the policy agenda (Keskitalo 2010a; Keskitalo et al. 2012a). However, relatively little attention was paid to the fact that municipalities – especially those with limited population, tax basis, and thereby economic resources – are not able to "mainstream" climate change or even strategic planning across their areas, resulting in highly varying implementation among municipalities (Keskitalo 2010a). To some extent, this can be seen as related to municipalities in the Swedish system being assumed to themselves generally manage risks that fall within the municipality itself, while the state should mainly act on issues that are beyond municipal scope. This understanding was highlighted in the discussions of what catastrophe management support the state should provide (cf. Keskitalo 2010a) and may to some extent have contributed to highly varying progress on adaptation in different municipalities. Municipalities have thus noted throughout that attention needs to be paid to the very different possibilities for municipalities to in fact act on adaptation and have criticised the limited government progress on adaptation while also noting the need for municipalities to themselves prioritise what actions to undertake on adaptation (Keskitalo 2010; Andersson et al. 2015).

The case of *Finland* is particularly relevant in comparison, as this was one of the first countries where a climate change adaptation strategy was formally decided on, and it can thus serve to contrast the slower development of formal adaptation policy in Sweden. However, the fast progress on adaptation in Finland seems to largely have been the consequence of adaptation being treated like any other issue, to be formalised quickly, according to the established working style (or logic, if you will). Once this had been done, however, implementation largely lagged (similar to Sweden, where formal policy development occurred later), as the specific resource and integration demands (not least on regions and municipalities concerning dealing with the adaptation issue) were not recognised, and scarce resources accompanied the formalisation of policy (Keskitalo 2010, ed).

In the *UK (England)*, which has often been seen as a forerunner in adaptation, adaptation was strongly supported through, for instance, being included in an indicator system with an impact on local council funding. This was largely seen as a model that was typical of the New Labour party approach to regulation. However, progress was constrained at the shift of governments when funding through this system was removed, and debate in accordance with a de-emphasis on adaptation at the time came to focus more on resilience (implying a greater focus on present-day issues such as the dredging of rivers rather than future risk) (cf. Keskitalo 2010b, pers. comm. in follow-up study, unpublished). Regardless of this, the institutionalisation of a climate law and follow-up frameworks including a focus and body targeting adaptation can be seen as a means by which adaptation has been able to remain a policy focus (i.e., one of the issues that policy-makers have to pay attention to).

Italy, finally, serves as an example opposite to that of the UK, as attempts at developing adaptation policy at the time did not even make it to being decided upon, largely due to formal requirements on the national level: A board with overarching decision prerogatives focused less on climate than on potential cost. Any work on adaptation at the time was thus voluntary at the regional and local levels and was not guided by policy (Keskitalo 2010, ed). (This study was not followed up after 2010, and it can thus not be commented on here how adaptation measures may have been integrated since then.)

The brief illustration shows that the role of existing regulative systems or organisation in how adaptation came to be managed was significant in all four countries.

The country cases can be – and were – understood through different theoretical lenses, which illustrates how multiple institutional issues affect how (and what) climate change adaptation measures are developed (or not).

In studies of the four cases, with a focus on *capacity*, Westerhoff et al. (2011) generally illustrated that multilevel adaptation to climate change depends particularly on institutional factors in terms of political leadership and distribution of responsibility for acting on adaptation, on both the national and local level. What municipalities can do is thereby determined both from above through allocated responsibility and to some extent resources, and at the locality to some extent in cases of resources (for instance tax base and size), specific human resources, and interlinkage into networks. It was difficult, however, to separate resources determined at the national level from those determined at the local level, as specific responsibilities and resources from the national level are related, for instance, to the size of the local authority (e.g. based on population) and could also be seen in relation to an interlinkage into networks or the ability to attract specific human resources. This study thereby also illustrated the need to see the local level in the context of higher levels, as well as to see resources in relation to their attribution and activation (cf. Bay-Larsen and Hovelsrud 2017). To some extent, this could also be seen as illustrating the limitations of a broad-capacity framework without clearly taking into account the activation of resources or how resources are gained or developed within a system.

A study of the same cases through the lens of *framing theory* (Juhola et al. 2011) illustrated that particularly the way weather events were perceived of as connected to climate change (or not), often related to whether there was a public discourse on climate in e.g. the media, was relevant to whether adaptation policies were seen as urgent. The way priority and resources were made available concerning climate change was thus largely formed through how the issue was framed, rather than merely in relation to underlying absolute "resources". Thus, the study further illustrated that adaptation can be seen as linked to different issue areas in different countries. For instance, in Sweden, adaptation was seen as linked to a vulnerability discourse and to internal political processes and existing support, which may have resulted in little direct shift in resources to municipalities. In the UK, on the contrary, adaptation was framed mainly in terms of both planning processes and economic risk (Juhola et al. 2011). These two framings together supplied both a venue for action (through the planning system) and an imperative or urgency, emphasising that action today would be economically rational compared with action in the future. In total, thus, the view of adaptation through a framing lens in this example highlighted which actors are to adapt, what they are to adapt to, how they are to adapt, and when they are to adapt, with the specific issue linkages of adaptation to being a rational economic issue in particular creating urgency. A framing focus could thus be seen as particularly highlighting how any resources are activated or even created within a system, in relation to what is portrayed as relevant in this system by the specific actors within it.

Whereas adaptation theory, particularly as related to adaptive capacities (Westerhoff et al. 2011), thus underlined the broad resources such as the institutional system that impacts adaptation, the framing perspective used in Juhola et al. (2011) illustrated in more detail the responsible actors within such a system and the extent to which they are also dependent on and made relevant through particular ways of perceiving and linking (even creating) the problem of climate change as a problem for certain actors. The study thereby highlights the ways in which political administrations and issues have to be enabled or attributed responsibility for certain actions, and how they, absent such an issue linkage, may not interact with an issue. The possibilities for these framings to gain effect may also be linked to the way a country is organised: In the UK, nationally developed aims can directly impact local and regional resource distribution, while the more decentralised state structures in Sweden and Finland imply a more limited ability of the state to directly steer municipalities in a way similar to that in the UK. These types of underlying structural issues may also masquerade as framings, for instance in Sweden and Finland, making issues become "framed as vulnerability to climate impacts at the local scale where extreme weather events have been felt" (Juhola et al. 2011: 459–460). In the UK, on the contrary, issues were framed not only as local but also expressively as national, as different actors turned to the national level for resources (the same way municipalities in Sweden have insisted that resource distribution issues need to be taken into account, Keskitalo 2010; Andersson et al. 2015).

This perspective of local-national interlinkages in terms of the importance of how an issue needs to be framed in order to be seen as a problem is further highlighted in Keskitalo et al. (2012a), which applied an *agenda-setting* perspective to the development of adaptation policy in the four countries. The agenda-setting perspective, largely drawing on Kingdon's study of in-practice policy-making in the US Congress and now widely applied, illustrates the way an issue progresses from being merely a situation to being seen as a problem and thus being on the "agenda" of issues to which policy-makers apply serious effort (Kingdon 1995). The theory assumes that at any point there are any number of issues on which policy-makers could work, but that time and resources are limited to those that are made urgent in some way. Issues are made urgent through changes in the political stream (for instance a change of the party in power), policy stream (for instance new reports), or problem stream (for instance extreme events or crises), which make it possible for policy entrepreneurs who have track records and interests within any particular issue to come to the fore and make their expertise relevant (Kingdon 1995). Keskitalo et al. (2012a) thus illustrate the great importance attributed to extreme events as crises that may lift the adaptation issue, if adaptation is framed this way. The study illustrated ways in which the UK policy, politics, and problem streams were highly favourable on multiple levels to the emergence of adaptation as an issue, as it was framed as a crisis that required responses at all levels, particularly in relation to recently experienced floods and flood risk. At lower levels, this understanding was also supported by, for instance, regional stakeholder bodies and voluntary local actions as well as developments in adaptation policy at local levels (Keskitalo et al. 2012a). In Sweden and Finland, on the contrary, the issue was formally treated mainly through actions on the national level that to some extent attributed the responsibility for action to the local level, although in the Swedish case it was forced by actions at lower levels; the more limited treatment could potentially be seen in relation to both a higher focus on mitigation and more limited extreme events (cf. Keskitalo 2010, ed).

Finally, a paper utilising a *governmentality* perspective (Keskitalo et al. 2012b) particularly illustrated the ways in which these diverging developments in the different states may be attributed not only to changes in different streams at the time of issue development but also to a broader national culture concerning how issues are dealt with. As noted earlier in this chapter, a governmentality framework can be applied to highlight the technologies of government (Rose 1996; Rose and Miller 1992) that may be used by different states with different levels of acceptance. The study (Keskitalo et al. 2012b) showed, for instance, that the UK system was supported in its multilevel treatment of adaptation not only by a framing and agenda-setting system that supported adaptation, but also by what actions were possible on a national to a local level in the UK system. Whereas Sweden and Finland have high self-government on the local level and may not accept strong steering by the state in issues conceived of as falling under local authority, the English administration under New Labour governed through a complex framework allowing nationally set indicators for local government performance evaluation, which also implicated local funding. At the time of the study, adaptation was included as one indicator on which local government performance was indicated; this made adaptation visible, measurable, and financially relevant for local authorities to focus on. It was also developed in accordance with the regulative style highlighted in the UK, particularly during the Blair years, illustrating the role of specific logics, sometimes wider than any one specific area or sector, in influencing, for instance, how climate change is governed.

The case, seen through a governmentality focus, thereby also demonstrates how specific knowledge and attributes were made visible and valued in relation to specific logics (similar to how specific logics in the Swedish forestry sector are applied to governing Swedish forestry, or specific logics in how resources are attributed in the Swedish system highlight some issues more than others) (cf. Keskitalo et al. 2012b).

Together, the different social theories used in this selection thereby illustrate a number of issues that add to a focus on capacities. These include the inherently political and complex social and socio-political nature of issue development and attribution of responsibility. They also indicate the ways in which different social theories – particularly through multiple studies applying multiple frameworks – may better serve to indicate the complexities of social systems and "barriers" to adaptation. In general, however, all these cases underscore the role of seeing adaptation as an institutional issue: How it is managed is perhaps determined more by the institutions that come into contact with it, or the effects of it, than by what knowledge about it exists "out there".

Understanding adaptation as an institutional issue

Using various cases and applications of theory, this chapter has illustrated how climate change manifests as an institutional issue. As climate change is not the only issue that different organisations, states, or levels need to deal with but is rather an emerging issue in the context of multiple already institutionalised issues and ways of working, already institutionalised issues and sectors, ways of working, and interests have to be understood in order to understand how they may act or might be made to act to act on climate change. The chapter has also illustrated that developing action on climate change adaptation and mitigation is far from automatic, even when actors are given information on climate change (and even among actors responsible for climate change adaptation, as they act and are delimited in how they can act within larger structures). The example of interviews with representatives holding adaptation roles in forest industry could perhaps be seen as illustrating this the most clearly: What interviewees in their organisations, even with formal roles involving adaptation, could do in this regard depended largely on how forestry was structured and what understandings were more generally forwarded on how forestry should work, what it should provide, and what issues were made relevant.

Thus, the empirical results highlight the importance of understanding not only the role of present institutions but also their past, and what interests they have formed in relation to, based on the knowledge that these interests often remain active and continue to guide development today. This is not least because organisations diverging too greatly from their constituencies may see the same consequences as in politics – waning support, or in this case, potential membership or economic loss. As a result, the studies drawn upon in this chapter show that incremental change, if any, involving what was nevertheless increasingly seen as an issue requiring greater transformation, was thereby largely the norm (see for instance Andersson and Keskitalo 2018; Keskitalo et al. 2016; Kunnas et al. 2019; Keskitalo and Pettersson 2012). Adaptation policy in different areas could functionally be seen as resulting from the existing policy orientation and interests encased in the sector.

This means that the present structure can also not be seen in isolation but must rather be understood in relation to the institutional environment in which it was formed and which it has also contributed to forming. Consequences may include some sectors or interests being better encased in legislation and able to forward their interests, while others may have less possibility to gain resources, including e.g. to adapt or mitigate in relation to other interests (e.g. Pettersson et al. 2017; Kunnas et al. 2019).

Thus, to understand the possibilities of gaining traction for change, it can be seen that not only the sectoral and different impacting areas must be understood – fields that often fall under such different legislative bodies that climate change issues are not possible to simply mainstream across them, as is often assumed in literature (Klein 2011; cf. Keskitalo and Pettersson 2016). Instead, sectors and actors must also be understood as part of the national – or even international – context in which they are enabled in specific ways. As a result, "rather than dismissing a focus on business-as-usual with regard to climate change as 'illogical' or 'incorrect', it is important to comprehend and explore the logics of adaptation that drive and structure these defined actions" (Andersson and Keskitalo 2018: 81).

Key points

- Forest in Sweden can be seen as a case of an institutionalised system that has developed over time. In this, it can be used to illustrate types of relations between organisations, state and private actors of different types, and interests that are relevant to understanding institutions and the logics in relation to which, for instance, policy instruments for supporting adaptation or mitigation could be developed.
- The case of forest in Sweden is impacted by multiple other processes, including international development, and illustrates the multilevel character of cases: that it may not be possible to inherently understand developments only on the local level or to understand the range of issues that drive the developments.

- Cases as well as a multilevel mechanisms can be understood through different theoretical frameworks. These can add an understanding not only of the case but also of the general types of mechanisms that are also likely to influence other cases (such as governmental logics or path dependency-related features).

Study questions

- What can we learn about institutions from discussing the case of forest in Sweden?
- Could you design a study – either a literature study or an empirical study – through which you would be able to gain this type of knowledge about a field you are interested in? (a country/county/local, sectoral, or other case)

Notes

1 Increased flooding may also result in nutrient leakage (particularly as forest is fertilised and fertiliser may leach) (Andersson and Keskitalo 2018).

2 Keskitalo and Pettersson (2012) also further note that, as mentioned in previous chapters (see Text Box 2.1 in Chapter 2, this volume), from a multilevel perspective there are additional regulations apart from, for instance, the limited formal competence of the EU involving forest that nonetheless impact forest. An example of this is the health risks posed to forest by invasive species; it has been noted that the maxim of free trade – the overall purpose of the World Trade Organization and also one of the central tenets of the EU – may be in conflict with emerging environmental principles. That is, there are limited possibilities to limit trade in products that may cause invasive species risks, such as potted plants, as it is not possible to target exact risks from this more general area without this entailing a prohibition of free trade under the WTO. This has resulted in limitations, e.g. in relation to invasive species risks, having been delimited to those that are possible to target in relation to scientific evidence on the particular species (cf. Box 2.1 in Chapter 2, this volume) (Keskitalo and Pettersson 2016). This regulative situation, despite recent developments in invasive species governance (Keskitalo and Pettersson 2016), can thus also be seen as limiting what actions can be taken to adapt to climate change in the forest context.

Additional readings

Keskitalo, E. C. H. and Pettersson, M. (2016). Can adaptation to climate change at all be mainstreamed in complex multi-level governance systems? A case study of forest-relevant policies at the EU and Swedish levels. In: Leal Filho, W., Adamson, K., Dunk, R. M., Azeiteiro, U. M., Illingworth, S. and Alves, F. (eds.) *Implementing Climate Change Adaptation in Cities and Communities. Integrating Strategies and Educational Approaches*. Springer, Dordrecht. Pp. 53–74.

5

WHY KNOWLEDGE IS NOT ENOUGH

Limits to communication and learning

Introduction

Previous chapters have problematised the assumptions of the linear model of expertise or scientific knowledge. They have also illustrated both what a focus on institutions may entail and how this can be exemplified in forest and multilevel cases.

What this and the following two chapters will do is to draw out the consequences of the focus on social science and the theoretical orientations highlighted previously. This will be done to more correctly understand the context that is relevant to terms that have been highlighted in relation to the linear model and climate change literature: knowledge and learning (this chapter), stakeholder participation (Chapter 6), and the multilevel context to the local or community level (Chapter 7).

Continuing the approach that a crucial venue for thinking about adaptation and mitigation has to be based on real-life, institutionally relevant analysis, one major focus in this and the following chapters, both implicitly and explicitly, is on power. In this understanding, "power" should imply the same as what we have understood from an institutional perspective: that what any person, organisation, or other body does must be understood within a social, economic, and political context (cf. Manicas 2006; Thornton et al. 2012; Barnett and Duvall 2005). As shown earlier in the book, institutions are historically developed in relation to power perspectives – what actors have been able to gain from specific developments, and also what developments may have benefitted central power as well as multiple actors, not least economically.

These types of empirically identified power perspectives – including the fact that actors (or whatever term one uses for different organisations, representatives, and even individuals) do what is workable for them in the situation, framing, and context in which they are placed – are a major absence in the linear model. The

DOI: 10.4324/9781003043867-5

linear model places knowledge in the centre and assumes that it is apolitical and given, and should therefore be of interest to all who come into contact with it (see Chapter 1).

However, as much social science work has expressed, if social relations are generally influenced by institutional dynamics, then knowledge must also be (cf. Manicas 2006; Thornton et al. 2012; Barnett and Duvall 2005). This does not necessarily mean that knowledge is "political", meaning that knowledge would be possible to reject solely on the basis of being institutionally formed. What it means is that knowledge is formed for use in specific contexts, may only be directly comprehensible to actors in these contexts, and – even when it is more broadly understandable to actors – may not be usable to them, of direct interest to them, or possible for them to forward action on in a decisive way compared to all other actions they may be required to prioritise. This is the social framing of knowledge (cf. Schön and Rein 1994): Its interpretation and use, at the very least, will be influenced by the institutional setting.

Extending into what this issue of recognising a power perspective can mean, by illustrating the very real consequences of applying different perspectives, this and the following chapters will illustrate the differences a social science understanding – one based on real-life analysis highlighting the broad basis of thinking that has been seen here as institutionally based – brings to these key elements. These understandings have the potential to change the way adaptation and mitigation are thought about, if only they come to replace the still influential, incorrect, linear model of scientific knowledge.

Understanding knowledge, learning, and power: introducing the Habermasian-Foucauldian debate

With regard to knowledge and learning, as noted earlier (Chapter 2), major research assessment work has largely highlighted both these concepts – despite that it is now recognised that better knowledge will not necessarily lead to change and that learning cannot be assumed to take place automatically or in all situations. To clarify why this is the case and delve more deeply into the factors that influence knowledge and learning (as well as participation and stakeholdership), the chapter will draw on one of the clearest discussions in literature on the role of knowledge and power: the debate between the *Foucauldian* and *Habermasian* perspectives.

While the two philosophers Jürgen Habermas and Michel Foucault are often regarded as the main figures of so-called critical theory, they developed their work on very different bases: Habermas has been seen as focusing on transcendental grounds for theory and as being more systematic theoretically and utopian politically, while Foucault has been seen as developing a more historically and empirically based interpretation (Poster 1984). Critics have noted that Habermas does not have a single empirical application in his entire main works, while Foucault bases his studies on historical empirical – real-life – situations and traces the thought systems and role of power in them (Poster 1984).

These two philosophers are particularly important not just because of the clear discussion between them in literature; they are also important because the earlier described assumptions regarding the role of knowledge and potential for knowledge transfer, learning, and participation that mark broader mainstream climate change and environmental research, and that relate to a linear model of scientific knowledge, can be said to lie squarely in the Habermasian camp (although this has seldom been recognised; Keskitalo and Preston 2019a; Ashenden and Owen 1999a eds; Oels 2019; Allen [A.] 2010).

To highlight in this chapter what may be seen as "philosophical" or "theoretical" perspectives thereby in no way suggests a lack of implications in reality. Whether one holds perspectives that are closer to what could be seen as a Habermasian or a Foucauldian perspective has major real-life implications, for instance on how one addresses issues of knowledge transferability or interaction with social groups, or even understands the social world. In fact, the two perspectives can be seen to relate to whether one focuses more on what has been seen here as real-life considerations – in which perspectives of power play a large role, as could be seen in the forest case in the previous chapter – or more on perspectives that may not be equally empirically grounded.

As noted earlier, assumptions in the linear model of scientific knowledge lie closer to a Habermasian camp. It is highly notable that these assumptions, in the way Habermas developed them, were seen by Habermas himself as applicable to an *ideal* speech situation – not one that describes how interaction or communication functions per se, but one that Habermas advances as an *ideal of how it should function* (Ashenden and Owen 1999a). Habermas then posits different *rules that would have to be adhered to* for this to happen; that is, for communication and interaction to function as if they were ideal. Habermas himself thereby acknowledged that his model was not made to approximate real life but instead to serve as a basis for how things could be made more ideal, if specific rules were adhered to. However, the rules he set up have largely been seen as utopian: It is not that interaction and communication might not function better if they were adhered to, but rather that the suggested rules are just not something people can be practically made to adhere to in real situations (e.g. Holmes and Scoones 2000; Rip 1986).

The Habermasian theory discussed in this chapter thereby differs from the institutional perspectives discussed previously, in that it is not based on analysing real-life situations and developing analytical tools for describing them, but rather on advancing an idea of how it should be. It is thereby a normative theory of how the world should be rather than how it is. This means that the perspective must inherently be seen as limited in relation to describing and analysing a real-world situation, which is what we try to do in this book (cf. Manicas 2006; Thornton et al. 2012).

The opposing example, which does focus on real-world, empirical, analysis is that of French philosopher Michel Foucault (e.g. Foucault 1974). Contrary to Habermas's ideal model, Foucault, who takes the real-world perspective, argues that all actual decision-making and development of knowledge is impacted by power and interest. In this understanding it is assumed that there are no interest-free "ideal" situations, and therefore the crucial point is to study the actual interests going into decision-making and their discourses and mentalities of governing (e.g. Foucault 1974). As we noted earlier, Foucauldian approaches are thereby akin to broader institutional approaches in the general aim to describe and analyse *actual* decision-making situations and some of the limitations to change in these situations (cf. Young 1999; Keskitalo et al. 2019; Mahoney and Thelen 2010).

This chapter thereby also illustrates several issues that are of importance in understanding social thinking.

Firstly, it illustrates the difficulty that not all theoretical perspectives are compatible, as they have been derived for different reasons and may entail different assumptions as to how the world works (or should work, or could be made to work). The consequences of taking one or the other perspective are resultantly large: Someone applying the linear model of scientific knowledge or a Habermasian perspective to the real world might assume that knowledge can be communicated directly to stakeholders, who will then learn and use it (particularly if they have not delved into the role of theory and the extent to which and under what circumstances which theory might be applicable). Someone applying a Foucauldian perspective – aware of the role of power and institutions in the real world – would instead also view knowledge, interaction, and communication with these eyes, assuming that any knowledge may require targeting those who can and want to use it but that it could still not be assumed to have direct impact, among other things given what other priorities various actors may have.

Secondly, given this background, the chapter also illustrates the importance of being aware that much literature is written by people in a specific field – for people in the same specific field, who are assumed to already know the basics of these perspective choices. For such reasons, single social science studies (or any single study) seldom explicitly describe the range of variations or limitations in compatibility of a theoretical perspective, or highlight the variations in assumptions regarding process (how things happen) applied by different perspectives. And actors who have been schooled in one approach, perhaps implicitly (such as the case of the linear model of scientific knowledge), may not even have reflected on the notion that these types of assumptions guide their thinking. The chapter can thereby also serve as a caution that all statements – including, for instance, assessment or programme texts – need to be read while applying this type of critical "meta" perspective: What perspectives does the text rest on – and am I sufficiently read up on the basic assumptions of these to assess the text and results in context? What evidence, if any, is provided that these perspectives are the most relevant ones to apply – or is it an example of a more unreflected, implicit assumption (such as the linear model)?

As this book is targeted towards firstly the climate change field, the seasoned philosopher will find the relatively superficial description of the Foucault-Habermas debate lacking: Neither Habermas's nor Foucault's major works are entirely unified (far from it, some would say); both authors also modify and explain their understandings over time, not least in relation to power (Ingram 2005). Thus, some of the debate sketched in this chapter also reflects some of the understandings of Foucault and Habermas in literature. Habermas, while in some understandings relating to assumptions similar to those in the linear model of scientific knowledge, cannot be reduced to only doing this (which is also why Habermas is highlighted here, as a counterpoint to Foucault, rather than earlier in the book) (e.g. Ingram 2005). However, the hope is that the chapter will at least serve to illustrate the difference between ideal and real life, and to contrast assumptions relatively absent a focus on power with those focusing on power in real-life situations, as relevant to the purpose in this book of problematising the types of framings used in climate change literature.

A Habermasian understanding

This specific topic of Habermasian-Foucauldian conflict of understandings has been the focus of much work in the social sciences (see, e.g., Ashenden and Owen 1999a eds; Mayes 2015 for outlines of the debate); thus, it is possible to go to these as well as to literature on each author to sketch the differences, not least as they have been expressed in relation to environmental studies.

As this book has shown throughout, assumptions forwarded not only in the climate change field but also in environmental studies far beyond it have often been seen as aligning with those in the linear model of scientific knowledge. In a chapter in a 2003 book called *Public Participation in Sustainability Science*, Ravetz summarises that participation here is largely assumed to be a process in which participants: a) express unvested, openly spoken (and genuine) perspectives without implications of power relations; and b) are able to learn from each other and internalise the perspectives that are spoken, and thus are able to understand arguments and take discussions to heart. This would assume that they: c) will implement their learnings or in some other way "make progress towards a sustainable world" (Ravetz 2003: 76). Although the linkage to Habermasian thinking has been less pronounced in this literature, some have noted that these factors are in coherence not only with the linear model of scientific knowledge but also with Habermas's analysis of the *ideal* participative process. Holmes and Scoones, for instance, in a paper on participatory environmental policy processes, argue that "deliberative inclusionary processes" including participation generally "appear to involve, at least implicitly, Habermas's ideas of 'communicative rationality'" (Holmes and Scoones 2000: 9). Others have noted that this may not be surprising, as Habermas early on intended to design programmes to contribute knowledge for decision-making about social priorities (Wuthnow et al. 1984) – a view, developed in coherence with broader rationalist research programmes of the time, that may have filtered into environmental decision-making. This makes it relevant to review in further detail Habermas's understanding of what this ideal is, but also to highlight the consequences of an ideal image in any way being assumed to be applicable to the real world.

In general, the ideal of "communicative rationality" advanced by Habermas is seen as a situation that involves free debate and dispute. In this situation, the only legitimate force is a good argument. Interactions are to be egalitarian and uncoerced, as well as free from deception, power, and strategy (e.g. Dryzek 1990; cf. Habermas 1984, 1987). The assumption is that a communication process, *if undertaken in this fashion*, would deliver the most "correct" political judgement possible, as it would lead to a consensus on values (Holmes and Scoones 2000).

From the outset, Habermas's ideal of communicative rationality was intended to counteract a sole focus on science and technology expertise, to also take into account moral obligations. It was thus expressively intended to offer all relevant actors a means of communication (Wuthnow et al. 1984). To develop such an ideal speech community, where all actors relevant to the public sphere would be able to meet, Habermas set up a number of rules for ideal communication including, for instance, abandoning interests and committing to consensus development. In this understanding, conflict and dissensus should be resolved through the notion that "only those norms of action are valid to which all possibly affected could assent as participants in rational discourses" (Habermas 1996: 459, 107; cf. Dean 1999). This would mean that all participants are assumed to make explicit and openly discuss specific means of action. It would also rest on the assumption that all participants relevant to the decision are present, can make these arguments explicit, and are not hindered by power perspectives, resources, or the like (e.g. Holmes and Scoones 2000).

However, while fewer have disputed that this type of conception could be an ideal, it has been largely noted that it may be so far from the real world as to be impossible to achieve. Eley states that Habermas's "conception of the public sphere amounts to an ideal of critical liberalism that remains historically unattained" (Eley 1992: 289). While one might suggest that participation would require multi-way communication between groups, consensually and non-hierarchically and under the assumption that everyone should be able to make clear and abide by specific rules for social interaction (e.g. Webler 1995), these are not characteristics of the political process or of any communication across power, as those between policy-makers and other stakeholders, or science and other stakeholders, inevitably are. A consequence of applying Habermasian norms in actual real-life situations – where they are not abided by – may thereby result in a corruption rather than idealisation of argumentation, as all relevant actors are never present, their possibilities to argue a case may differ, and people never leave their interests at the door (Holmes and Scoones 2000). It has also been argued that, in emphasising the rational and cognitive dimensions of communication, Habermas also pays little attention to the fact that feelings and values are not always made explicit and may not be possible to change through rational argumentation (e.g. Wuthnow et al. 1984).

Habermas has thereby been criticised for neglecting the fact that certain practices and understandings are only understandable on the basis of tradition and history, which are seldom reflected upon explicitly and clearly (as we have seen evidence of in previous chapters). It has also been noted that Habermas largely ignores the fundamental institutions of socialisation such as the family, schools, law, or power relations whereby people are taught how to behave and follow specific social norms, sometimes in a way that makes them unquestioned and given rather than explicit (Wuthnow et al. 1984).

Habermas's conception is thus "ideal" in the meaning that it reflects how communication should be, rather than what it is. As a result, Habermasian notions of communicative rationality have been strongly criticised for a naivety concerning actual power relations. Rip suggests: "It is not realistic to expect the parties in a controversial issue to stop their strategizing, sit down together for a *herrschaftsfreie Diskussion* (Jürgen Habermas's ideal of a dominance-free debate), and reinsert the results into the struggle so that it will be resolved" (Rip 1986: 362). As Holmes and Scoones note, groups of stakeholders, instead of acting for any general public good, will form pacts to ensure that their viewpoint succeeds. To this end they may misrepresent themselves, stay quiet, or take on positions they do not hold. This will happen even if they commit to being open and honest in the debate. "Communicative action is therefore inherently political and powerful, as it is unable to control the individual through processes of stakeholders or to guarantee that all participants will act in an open and honest manner all the time" (Holmes and Scoones 2000: 32). Instead of acting in the prescribed way, groups may "see participation as providing new openings . . . in which to pursue their own agenda and the opportunity to launch a sustained conflict-style strategy within a collaborative framework" (Holmes and Scoones 2000: 32; cf. Hadden 1995).

Thus, to simply assume or even mandate that interaction should be rational and open would, in real life, not be a workable option. Rather than allowing parties to argue openly, "creating enclaves of relative peace only means that what happens there will be irrelevant to the battle (or the products will be picked to pieces as soon as they are re-introduced into the arena)" (Rip 1986: 362). Thus, Rip argues, similar to what has been emphasised throughout this volume, that:

> robustness cannot be created by stepping outside the arena, isolating the debate from social alignments, or neutralizing them procedurally (for instance, through representation of "all" interests and viewpoints), as is often proposed. *One has to work with, and through, the strategizing of the actors.*
>
> *(Rip 1986: 362, my emphasis)*

To work with, or through, the strategising of actors would then mean understanding, for instance, what knowledge people are receptive to, or even what incentives can be provided (as discussed in previous chapters) rather than assuming that knowledge exchange or learning will necessarily take place.

In fact, the understanding of learning that is found in the linear model of scientific knowledge, and that is related to how people take on new arguments in this type of interaction process, can also be related to Habermas. In his model, Habermas assumes that learning occurs in a series of irreversible, inevitable, discrete, and increasingly complex stages of development. As higher stages of development are eventually reached, the individual is assumed to become more autonomous, gain personal independence, become more capable of solving problems, and become more "rational" and competent as a "speaker" or social actor (Wuthnow et al. 1984: 201). This conception would imply that there is a scale of progression and that there are better, not just different, ways of understanding. It may also assume that there is some automation involved and that there is only one type of rationality: that learning is both inevitable and leads to more "rational" results.

These considerations have also been criticised from a focus on how things work in real life. It has been noted that real-life empirical studies show that there are multiple institutional influences and structures that impact what will be learned or used, and that it is not always possible to grade social conflict issues in terms of different understandings being "better", "worse", "wrong", "right", or more or less "rational". Confronting people with well-reasoned persuasive messages may often have little impact on their attitudes, especially if these attitudes are strong (Tenbrunsel et al. 1997). In fact, research on resistance to attitude change shows that it is more difficult to change attitudes the stronger they are – that is, when they are embedded in a knowledge structure involving issues on an intra-attitudinal or inter-attitudinal basis, meaning that they are supported by many associations arising from direct and indirect experience. Certain attitudes may also be linked to other attitudes by inference and can particularly be linked to more abstract attitudes, including values (Eagly and Kulesa 1997). Attitude change is thereby made difficult because attitudes are interconnected: When one attitude is changed, there are psychological reverberations for others (Tenbrunsel et al. 1997). This could be seen as somewhat similar to the argument that different "paths" or "logics" delimit what we focus on and learn, and what decision-making takes place (for instance, resulting in lock-ins), as earlier chapters have illustrated. In some relation to this, authors have also argued that there is a possibility that any learning that does take place is inert; that is, it results in the failure to use new learning outside the immediate context in which it has been learned (Gentner and Whitley 1997). Learning is also seldom translated into action: Research on the attitude-behaviour relation shows that simple expectations regarding strong attitude-behaviour relations will typically be disconfirmed (Eagly and Kulesa 1997).

As a result of this, behaviours that are detrimental to the environment are often not the direct result of people being indifferent to the environment. On the contrary, people may often act in ways that are harmful to the environment *despite* holding attitudes that are environmentally friendly or having learned information about the role of the environment. This is because trade-offs in terms of direct cost to the individual may exist, and may take place, for instance, in an action that is not directly connected to an environmental issue for the individual (for instance in a purchase situation). What happens, for instance, may be that the individual assesses the purchase in relation to its immediate use for themselves, rather than in relation to their environmental values (Tenbrunsel et al. 1997; Slovic 1997).

Finally, it is also rather commonplace that individuals or groups hold different understandings, whereby one is not better than the other but whereby they are determined by their interests. Habermas avoids this issue and is able to create a ranking of understanding by not allowing participants to have interests. However, in doing so, he also moves away from how the social world works and from the fact that people have interests – are "stakeholders" – whether or not they are procedurally permitted to (cf. Holmes and Scoones 2000).

While broader social science research thereby does not say that learning does not take place, it does problematise that it can be assumed to happen as a result of short-term, direct interaction that is removed from the considerations of an actor's everyday life (Rip 1986). The real-life characteristics of communication thus pose problems both for the attainment of Habermas's ideal speech situation and for any consultation design based on similar assumptions (as will be discussed in the following chapters).

A Foucauldian understanding

One way a problematisation of the communication process could be developed would be to understand some of the arguments and critique forwarded in regard to Habermas's model. As Foucault has often been seen as the direct counterpoint to Habermas, the following sections describe this approach – critical of, and often constructed and explained in relation to, Habermas in literature (as an intended debate between the two could not be undertaken due to Foucault's passing; cf. Foucault 1973, 1974; Mayes 2015).

Among other things, Foucault establishes the idea of discourse, as a power relation expressed in social interaction and language. This is something we have already briefly seen exemplified earlier in the book. One of Foucault's main arguments is that power is always present and acts upon what people can think, say, and do, even in situations that do not seem violent or forcing. This means that people are always positioned by existing discourses – that is, affected by power dynamics – within which they act, which may for instance make certain things possible and other things practically impossible to argue in specific settings. Foucault thus defines discourse as "constituted by all that was said in all statements that named it, divided it up, described it, explained it, traced its developments, indicated its various correlations, judged it, and possibly gave it speech by articulating its name, discourses that were taken as its own" (Foucault 1974: 32).

This point of departure means that nothing resembling an ideal speech situation exists for Foucault: Power is always present in people's interests, in affecting them in relations to authority, and in language and what can be said and what cannot. There would also not exist anything like unhindered communication, as all communication is per definition always hindered, unfree, in that it is influenced by one's previous understandings and the ways of understanding the world that one is schooled in (e.g. Allen [B.] 2010).[1] On the other hand, this situatedness is also what enables interaction: One is only able to speak from what one has naturalised and from the position in which one is.

Foucault developed his discourse analysis to highlight the system of rules through which power is legitimated: The way he considers discourse analysis to function involves identifying the limits of a given discourse through what is included and what is excluded in statements (Foucault 1974). This means that discourse is not only the study of language but also that of the connection of language (what can be said) to what is obscured, as a result of the naturalisation of a manner and assumption of speaking. Discourse analysis would thus target mechanisms of inclusion and exclusion: what is left out, and also, who are the speakers in terms of, for instance, disciplinary background, organisational position, and the like (Foucault 1974)?[2]

What one can do in order to work based on an understanding of power, then, is to trace the systems and features according to which communication and actors are hindered or even constructed in each situation (for instance, as briefly illustrated in previous chapters). Foucault himself focused mainly on contrasting features of major systems of thought, often over historically long times, but would certainly not have found it inapplicable to study specific lower-level cases. As Foucault scholar Rouse suggests in an illustration of how theory can be used, the mechanisms may be similar on any level:

> Epistemic conflict [conflicts on understandings of knowledge] is always shaped by the goods, practices and projects whose allocation and pursuit are at issue, and by the institutions and social networks that are organised around those pursuits. In such real contexts, there are constraints upon which arguments and which evidence will count as relevant and persuasive, based upon the need for support from others and for reliability from things. It matters what will count as persuasive to others who occupy strategic points in the circulation of knowledge and argument, and it also matters how things will manifest themselves in the contexts in which their behavior is recognized to be relevant.
>
> *(Rouse 1996: 413)*

In this statement, Rouse illustrates that power is always present, and that any real-life contexts could be assumed to include these types of constraints around what are the relevant (possible to make) arguments and recognisable evidence. Similarly, Rouse has noted that this recognition – that power and institutional perspectives are always present – does not mean that the results of any political or other process are necessarily discounted. Instead, they need to be qualified – in fact, in the same way as this book qualifies, for instance, results from assessment processes: While their description of the issues they focus on may be correct, they cannot be assumed to provide the full picture of, for instance, social relations or social context. Rouse states:

> Epistemic [knowledge] contexts are always in flux; their boundaries and configuration are continually challenged and partially reconstructed, as epistemic alignments shift. And these alignments are always intertwined with alignments of power and political resistance. To recognize this interconnection is not to devalue knowledge or science for political purposes, but to take seriously the stakes in struggles for knowledge and truth, and to place epistemology and philosophy of science squarely in their midst.
>
> *(Rouse 1996: 416)*

The perspective thus explicitly argues that discourses established in this way constitute the regular means by which the world is delimited and made known: Knowledge is established in the context of, and to support, power, as described here as ever-present and as part of forming the social setting. This means that what any groups concerned regard as knowledge "is that of which one can speak in a discursive practice" (Foucault 1974; 182).

In a Foucauldian understanding, then, knowledge is related to power. People are not able to shrug off their "interests", but are in fact *constituted* by where they are and how they have come to be there. This means that they will in fact not be able to be changed and "learn" in any unproblematic sense, even in deliberative settings or when confronted with evidence.

As a result, expressed in a simplified way, Foucault assumes that people are so stuck in their own understandings and ways of seeing the world (what he calls discourses) that they are in fact not free individuals, for instance unaligned and unitary "rational" "economic actors", who would only make choices based on economic rationale or what they have agreed to in a given situation. Instead, Foucault argues that one can see individuals as "subject positions". He makes this designation in order to, among other things, clearly point out that what any person identifies as important will depend on how they are situated in a context as a subject, rather than on any external "rational" argument (Dreyfus and Rabinow 2014).

In Foucault's view, then, contrary to Habermas, the notion that consensus would always be achievable would be based on a fundamental misunderstanding of power relations as purely negative – as if people would be pure and agree were it not for their interests, which in some way would "contaminate" them (McNay 1994). This idea that interests, or the messiness of the social world, would "contaminate" evidence can also be seen in the linear model of scientific knowledge, and as one reason why the "social" and even social study have largely been exempted from "science" in this understanding. However, as we have noted, this has also meant that "science" has been left incomplete and resting on incorrect assumptions of the social with no means to correct these assumptions and contribute to actual impact in the social system, as through this it also omits social study. Contrary to this, rather than in any way exempting the social world and the roles of interests in it, Foucault instead argues that we are where we stand and sit. In this type of understanding, we are constituted by the sum of our experience and thereby our interests, and as such cannot even envision a state outside them or outside power (or what has been regarded in this volume as institutions).

This does not mean that we are in any way "contaminated" by our interests, as if this could then be removed, but rather that we should recognise that this is how it is. Foucault states:

> The thought that there could be a state of communication which would be such that the games of truth could circulate freely, without obstacles, without constraint, and without coercive effects, seems to me to be Utopia. It is being blind to the fact that relations of power are not something bad in themselves,

> from which one must free one's self. I don't believe there can be a society without relations of power.
>
> *(Foucault 1988: 18, quoted in McNay 1994: 125)*

This does not mean that Foucault does not believe there cannot be consensus or agreement (McNay 1994). What it does mean is that people cannot be assumed to agree and reach a "right" result, as Habermas might argue, even if they are faced with the same knowledge. Perhaps for these reasons, Foucault also highlighted the need to always attempt to include the understandings of those who do not make it to the discussion table (or gain a voice through the "framings" there; cf. Schön and Rein 1994). One of Foucault's main approaches, genealogy, or the way the origin of discourses is traced historically, has been explained as:

> research directed towards a resurrection of local, popular, and disqualified knowledges through the production of critical discourses . . . to disrupt those forms of knowledge which have assumed a self-evident quality. . . . Therefore, at the very heart of genealogical analysis, is the activity of critique, rather than, for example, the provision of programmes, prophecies, or policies.
>
> *(Smart 1983: 135)*

This would mean that an understanding of communication through this historically developed basis would centre on actors' understandings. For instance, it would expect and try to describe patterns of dissensus, rather than seeing them as problems on the way to a necessary consensus. Most of all, it would try to reach individuals' or groups' understandings of their situation, in relation to, for instance, predominant ways of seeing an issue. In an application of this, rather than emphasising the fact that stakeholders do not prioritise science as a problem, one might try to describe their network of priorities. In an institutionally focused understanding, one might also then be able to discuss types of incentives or instruments that can be applied in such systems to steer actors (as noted earlier, this was not a focus of Foucault's, but is nevertheless something that developing an understanding of actual power relations could support the development of) (see Text Box 5.1).

TEXT BOX 5.1 AN EXAMPLE OF FOREST DISCOURSES

The previous chapter drew upon examples from the forest case and provided a short application of a Foucauldian governmentality perspective. A Foucauldian discourse perspective can also be placed on forest, in order to understand not only the various – and to the respective participants, valid – discourses on forest but also why they can be difficult to reach consensus on.

In line with a Foucauldian long-term analysis, authors such as Lisberg Jensen have drawn a parallel between the *production* focus in forestry (based on the Swedish case) and the modernisation project of the twentieth century. Modernisation ideas can be seen as based in Enlightenment and anthropocentric ideals such as control over nature and large-scale industrial rationalisation (Lisberg Jensen 2002). Lisberg Jensen sees the production perspective as contrasted by a *protection*-focused green environmental discourse. While this relates more to Romanticist ideas of nature as pristine and pure, she notes that this discourse has developed its own understanding of rationality in relation to modernist or Enlightenment conceptions, but here related to environmental science with a focus on conservation (Lisberg Jensen 2002). She thus notes that the ways people argue about the environment are based in historically developed understandings that may steer how they think today.

Given these different conceptions of the world and knowledge that production and protection perspectives may hold, it may follow that the groups drawing on different understandings also see forestry in highly different ways and find it difficult to meet in a consensus. The risk may be that any "consensus" that is developed (for instance, by forest production and environmental NGOs regarding certification) will conceal large underlying differences. This type of understanding, while it can be procedurally described in a few sentences (as in this text box), may in the lived reality of participants make a discussion of underlying aims and assumptions very difficult.

In a study illustrating the continuous importance of the two discourses in Sweden, it could be seen that the focus on, for instance, making good on the 1993 Forest Act aims of production and protection as equal targets has taken highly different forms, despite the fact that neither side formally contests this formulation (Lazlo Ambjörnsson et al. 2016). The forest industry has highlighted, for instance, intensive forestry and growth-increasing measures, and noted the large shifts that have been voluntarily undertaken through certification as a market-based framework (Keskitalo and Liljenfeldt 2014). Environmental NGOs take different perspectives on certification: Arguing that production has far to go to reach the international United Nations Nagoya target of 17% protected land area, some are more positive to discussion, while there are examples of NGOs leaving the certification system as they do not find it sufficiently far-reaching (Lazlo Ambjörnsson et al. 2016; cf. Cashore 2004). "Bridging" concepts and aiming to connect production and protection may thus not necessarily reveal only consensus but also dissensus within the development of instruments. Lazlo Ambjörnsson et al. (2016) conclude that:

> "Although green values, such as biodiversity, have been incorporated in the modern production discourse, and the green environmental

discourse has incorporated ideals of modernity, there is a major discrepancy concerning focus and meaning, even though terminology is converging."

(Lazlo Ambjörnsson et al. 2016: 116)

The development of systems for including or shifting aims – which may even have developed to provide a perspective around which actors can unite without changing their underlying assumptions – may thus hide large and continued variations – and conflict – in the focus on and understanding of, for instance, resource use.

What does the difference between Habermasian and Foucauldian understandings mean for an understanding of change, adaptation, and mitigation?

As shown in the chapter on theory and institutions, change is difficult. The Foucauldian perspective on the world aligns more with the perspective on institutions than the Habermasian one does, simply because Foucault deals with the real world – how things actually are – just like a broad institutional focus does, while Habermas deals with the world as he would like it to be: one of ideal communication.

The previous sections have presented the Habermasian ideal, not as just any problem with a philosophical perspective – as it may indeed be useful to think about how the world could be – but as a major problem when it comes to applying such an ideal-type perspective to the actual world, by assuming that the world can be made to function as if it were "ideal".

As a result, one might assume that both communication and knowledge transfer and a number of related conceptualisations regarding stakeholder participation, learning, and change in relation to these would be much harder to develop than elements in an ideal type or related linear model of scientific knowledge would assume.

Resultantly, while Foucault does not assert that change is impossible, much more than in a Habermasian understanding is required for change to take place. A Foucauldian perspective might imply that people are unlikely to understand arguments for change unless they are spoken and manifested by actors who are also schooled in their specific discourse and thus able to phrase and manifest requirements and examples for change in terms they both understand and perhaps even become forced to respond to (as it might otherwise even undermine their position). The concept of power is integral here: Power and knowledge are seen by Foucault as co-constitutive, and actors who express specific discourses thus do so in terms of both the power and the knowledge that situate them (Allen [B] 2010).

This conception means that, in general, people cannot be assumed to understand problems far removed from themselves in a way that greatly motivates them to attend to them (Allen [A.] 2010). In fact, they may not even recognise or agree on the problem description, or framing. In the terms of other theories used in social sciences, discussed earlier (Chapter 3), situations can thus be seen to potentially exhibit path-dependent features perhaps akin to the features Foucault traces through time as making up discourses. In this way, if central elements of these ways of seeing and being in the world are threatened, it would require a great deal to move from, for instance, incremental to catalytic change (or transformation).[3] For organisations, actors, or other units in which people play a role – that is, the whole social world – any knowledge and action in relation to it would thus need to be motivated in regard to their social situation and aims for it to be made relevant to them. Examples of this can be that, within the decision-making horizon of a business, a change would need to severely impact their bottom line, for instance the willingness of investors to invest. For investors, it may include that their decision-making horizon is so long – or that changes like climate change are becoming so urgent – that they start targeting investment in businesses that may do well under climate change and withdrawing from those that may not.

As a result, it may thus not be surprising that decisive action on climate change, or the environment, is taking a relatively long time. However, we also have to recognise the changes that have been made, and how they may have provided the foundations for changes to come. We also need to recognise that, when perspectives have been forwarded, this may not always have come from within organisations but has rather been due to outside pressure. In a Foucauldian perspective, there are entry points into discourse and power, but they are delimited: Foucault argues, similar to some institutionalist perspectives, that change often comes from outside the discourse and thereby from the perspectives of people or organisations that are not naturalised into a dominant perspective (similar to, for instance, agenda-setting perspectives, in which external shocks are prominent as changes that place a new issue on the agenda or that could, for instance, be said to break path dependency) (Ashenden and Owen 1999b; Allen [A.] 2010; cf. North 1990; Kingdon 1995).

So what does this say about adaptation, or about action on climate change or environmental issues in general? Taking the extremely broad birds'-eye view on social theory that this book does, comparing theories that those schooled in each theory would see great differences between beyond this birds'-eye view, one might argue that the implication remains that *systems cannot be assumed to shift easily* (as noted in Chapter 3). What this mainly means for arguments concerning learning, communication, interaction, and mainstreaming is that none of these processes of intended impact on stakeholder or policy processes can be seen as automatic or be assumed, and that "mainstreaming" (of, for instance, climate or environmental concerns into policies at large) cannot be taken for granted. Instead, attention must be paid to the institutional context of what measures, information, or instruments can be made to play a role. At the very least, the arguments here show that, rather than assuming that knowledge is enough and that learning will take place, we might be better off reviewing each step of what would need to happen to get a certain policy to a certain place through a scope of "Where can it go wrong?" (rather than assuming that it will go right). So, if something does not happen, it may be because the trade-offs to changing something are not seen as sufficiently politically valuable in the current situation, rather than because people act "wrongly" and could be made to act "correctly" merely through persuasion. As people, in the real world, may generally act within their immediate interests and conceptions of a situation, a way to develop change thus remains to incentivise specific actions – change what the costs to actors are – e.g. through reviewing potential economic or other instruments that can be applied, based on an understanding of the motivations and institutional dynamics that exist.

Starting from this kind of understanding of social interaction, one would place the emphasis not on making communication rational but rather on ensuring that people, individually and otherwise, are able to clearly articulate their "irrationalities", or rather, their rationalities as formed and defined by their everyday constraints and considerations. It should also not be assumed that it is possible to directly be able to understand, access, and change people's social worlds; perhaps a more modest and exact approach should be taken, starting with understanding them. Foremost, it is important to problematise the assumptions behind communication and how any "ideal" situations may differ from other, more empirically based, understandings of processes of interaction.

Key points

- The understanding of knowledge inherent in the linear model of scientific knowledge can be seen as linked to an understanding of knowledge related to ideal situations.
- Habermasian understandings of knowledge and related processes can be seen as reflected in many environmental science assumptions. They thus have significant consequences on assumptions regarding, for instance, learning, participation, and interaction with stakeholders.
- Habermasian perspective can be contrasted with Foucauldian perspective, which highlights the role of power in all interactions. It highlights that no interaction can be interest-free or free of power. In this, it has a pattern resemblance to understandings of institutions as formed over time in relation to power, and it acknowledges real-world situations relevant to communication and decision-making.
- Contrasting these perspectives can help us identify different perspectives more clearly, and also understand the implications of different assumptions for adaptation and mitigation in the real world, such as the fact that mainstreaming (or any other process intended to influence decision-making) cannot be assumed to proceed automatically based on identified knowledge.

Study questions

- Why is a comparison between Habermasian and Foucauldian perspectives relevant to an understanding of climate change and environmental studies?
- Why is a comparison between Habermasian and Foucauldian perspectives relevant to a focus on institutions?
- What can we understand about institutions based on a discussion of power?

Notes

1 This is an explicit point recognised by Foucault; however, on Habermasian thought it has also been noted that even if it attempts to enable "free" and "unhindered" discussion, the way it is intended to be undertaken is in fact related to specific values and would need to be learned: To this extent, even a Habermasian ideal situation would be "unfree" as it requires and regulates specific means of acting that may be more easy to gain power within for certain groups in society, notably those who are already well powered through, for instance, existing discourses making it possible to voice their arguments in a clear way (Kulynych 1997; cf. Oels 2019).

2 Such a discourse analysis, like any thematic analysis, is typically generally undertaken through text review, as language is the way discourses manifest most clearly and can be found in analysable form (although text can be drawn from multiple sources, including transcribed interviews or speech) (Keskitalo 2004).

3 On the other hand, some of the criticism levelled against both Foucault and path dependency (in very different contexts and terms) is also valid: It is difficult to identify what can or cannot be changed until afterwards, or without identification through genealogical and archaeological discourse study means, which require some history of a development, whereby patterns may only be clearly identifiable post-development (e.g. Gorges 2001).

Additional readings

Ashenden, S. and Owen, D. (1999) Introduction: Foucault, Habermas and the Politics of Critique. In: Ashenden, S. and Owen, D. (eds.) *Foucault Contra Habermas: Recasting the Dialogue between Genealogy and Critical Theory*. Sage Publications, London.

6

WHY UNDERSTANDING STAKEHOLDER PARTICIPATION REQUIRES UNDERSTANDING POWER AND INSTITUTIONS

Introduction

To understand society, economy, and politics, the argument in this book has been that one has to understand institutional dynamics. And, as the previous chapter showed, institutions are not formed in nor do they exist in a vacuum but are rather formed in and by relations of power: power among actors, organisations, perspectives, and framings, sometimes over centuries and providing traces or paths into the present. This type of understanding, which means that power is not only a resource or something one has or does not have, but instead something that is omnipresent and impacts how we can act, is crucial to understanding what role knowledge can play, and for whom (Barnett and Duvall 2005).

The previous chapter showed that what we see as knowledge can itself be seen as constituted within this relation between groups and for groups (cf. Beck 2011). Much of the understanding in this chapter follows relatively directly from the problematisation developed in the previous one. What the present chapter does is draw out the conclusions of a conceptualisation of knowledge and learning based in real-life constraints, for stakeholder participation. The chapter also provides examples of the ways in which a more limited conceptualisation of stakeholder interaction can be found in climate change-related literature.

Thus, the chapter illustrates that, similar to many other considerations in the climate change field, a conceptualisation of "stakeholders" and stakeholder participation processes has been subject to a limited understanding of institutional real-life processes and has been related more to the linear model of scientific knowledge, or Habermasian assumptions. Contrary to ignoring or taking as given that a process will lead to change, or that stakeholder participation will inherently lead to fairness, this chapter will instead highlight the issues regarding power and institutional perspectives that influence stakeholder participation or direct interaction and how it may occur in actual adaptation and mitigation, drawing broadly on social literature.

DOI: 10.4324/9781003043867-6

What is stakeholder participation?

In modern environmental policy, the rhetoric of public participation in the decision-making process has come to play an important role. The role and practice of participation, however, are more difficult to define because participation is taken to cover a large area with varying definitions depending on the case. In its broadest form, participation can include areas from education and information to interaction and dialogue. Often organised by government or science, it can also be organised by other groups and can be seen to range from lobbying or protests to public hearings and formal means of interaction with decision-making bodies (Spyke 1999; cf. Pettersson et al. 2017).

In general, the approach of public participation has been derived from a philosophy to supplement the political process. In this understanding, participation draws its roots from *participation theory*, anchored by the democratic values stating that, because government is derived from the people, all citizens have the right to influence governmental decisions, and the government should respond to them (Spyke 1999). Such a view, derived from "classical" theorists of participatory democracy such as Rousseau and John Stuart Mill, focuses on the relation of government to its constituency (Pateman 1970).

However, more recently this ideal of involving broader groups than those directly responsible for making a certain decision has been applied not only in policy but also in business and in relation to scientific prioritisation. One has thus come to speak of *stakeholders*, generally seen as those individuals or socially organised groups who are affected by – "have a stake in" – a certain issue or decision, whatever authoritative body and field this may be related to (Van de Kerkhof and Wieczorek 2003; cf. Keskitalo 2004). Motives for involving stakeholders include a desire to increase their awareness and acceptance of measures, and thereby the effectiveness of decisions or prioritisations. Stakeholder involvement also can be seen as a way to increase the legitimacy of decisions and enhance their accountability. Stakeholders are also seen as potentially being able to provide additional information or viewpoints to scientists or policy-makers, which may lead to better-informed decisions (Van de Kerkhof and Wieczorek 2003; Beierle 2002). In addition, as we have noted in Chapter 2, including stakeholders in participation in relation to scientific aims has been a crucial way in which the linear model of scientific knowledge has been modified in response to requirements for including arguments beyond scientific ones.

However, whatever the organising body, the aims of those organising the participation often differ from those invited to participate. Politicians, science, and business often have abstract or high-level goals removed from the immediate concerns of constituents and may view participation programmes as a way to improve their already organised and institutionally coloured decision-making process. To achieve this end, they may strive to exchange information with diverse groups within the target community, demonstrate a responsiveness to their concerns, and ultimately gain acceptance of their decisions (Spyke 1999). Participants, on the other hand, are often motivated by immediate goals: On average, people take part mainly because of their interest in the practical outcomes of the processes and how these may influence their everyday lives. Their support for these outcomes is crucially dependent on how they perceive quality in terms of their own immediate or short-term preferences (Enserink and Monnikhof 2003).

Because of this wide variety in stakeholder processes, and the often limited problematisation of how communication across these fields of experience should take place, discussions on the substantive issues may be complicated. Inherent problems include the possibility that the emphasis on the direct access to decision-makers in public participation may provide special access to interest groups outside the representative system of elected bodies and decision-makers. This type of access may result in a more prominent role for specific interests or individuals (e.g. Spyke 1999). For such reasons, "more participation" does not necessarily equal "more democracy", as it may support already powerful or special interests rather than the broader cross-section of interests that exist in relation to an issue. This is also connected to the problem that there exist no criteria for the objective selection of stakeholders. As only a selection of stakeholders is typically involved, this raises inherent questions about representativeness and selectivity, including mechanisms for self-selection (Van de Kerkhof and Wieczorek 2003; Holmes and Scoones 2000).

There are also related problems with participation programmes from the public's perspective. Whoever does participate is likely to experience a drain in terms of time and personal cost, for example the time and possibly frustration involved in becoming comfortable with the technical nature of many issues, even in a well-developed participatory process. This is an issue that again makes it more likely that mainly well-resourced and often adversarial groups – groups that have a well-developed agenda and established conflicts with other groups – commit to the processes, in order to gain influence. Interest groups, public or otherwise, may at times come to take over the agenda (Spyke 1999). Because of the difficulty involved in identifying and involving "the general public", participation thus often comes to be directed towards sectoral and organised actors despite the rhetoric of citizens, civil society, people, the general public, and so on (Magnette 2003).

In the organising body's perspective, participation may also undermine any administrative goals of efficiency, expertise, and control, which may drive agencies to seek quick public approval of predetermined solutions. On a practical level, public participation requires a great deal of both funding and time. Public participation can also result in lowest-common-denominator solutions if decision-makers strive to accommodate as many views as possible (Spyke 1999).

On a more practical note, well-developed participation may also be out of scope given time and resource limitations – even in cases in which it may be of use or even required by various agencies. There are practical limitations in relation to time and resources, as well as possibilities to design processes in relation to often broad and varying scopes of stakeholders, that add to process limitations. Thus, some agencies or organisational actors may not be consistent in how they approach public participation, may even fail to plan for it, or experience difficulty coordinating participation efforts with other agencies involved in the same or similar projects. Agencies or actors may furthermore have few determined output channels in the form of impact on decisions or the implementation of the needs indicated in participation (Spyke 1999).

Difficulties like these may practically mean that, even if stakeholders are involved, there may not be clear means developed for taking into account their suggestions; suggestions may be gathered in multiple and not necessarily interlinked processes on similar issues; and what stakeholders are invited to take part in may differ or not be thought through, for instance as a result of time and resource limitations on the part of the organising agency.

Participation at different levels of organisation

Participation also includes the question of the level at which the process and participation take place (an issue that will be further problematised in the next chapter). These levels may range from local to regional and national or even international, in the way we have earlier conceived of multilevel governance systems (e.g. Chapters 3 and 4). Varying levels can all be seen to entail specific considerations in the design of stakeholder processes, and interaction may also be complicated by participants acting on several of these by "jumping levels" (for instance, acting on several levels and in several venues to forward an agenda; cf. Gupta 2008).

On a local level, stakeholder processes have perhaps been practised longer and are often more straightforward, often concerning practical foci such as a waste treatment plant or the localisation of a new industry. This may seem to provide for a more straightforward selection of participants as, for instance, those interest organisations and citizens who are directly affected may provide a relatively clear motivation for participants to attend. On the local level, participation processes are also generally recognised through, among other things, the protection of democracy, basic civil and political rights, and people's development, economic, and environmental rights. Key rationales for participation here include, for instance, the view that local community democracy involves more than elections and requires meaningful dialogue, debate, and discussion in efforts to solve problems that may arise in the community (Sisk et al. 2001). However, even if the selection or self-selection of participants may be more straightforward at the local level than at other levels – in that those who are directly influenced by, for instance, a road development may participate in meetings about it – it is not a given that local participation will be able to take in all the "stakes". For instance, not all potential participants with local stakes may be residents or available locally. Any "local" participant can also not be assumed to cover all "local" perspectives, as there may be significant division among groups and individuals even locally. It is also difficult to determine what the "limits" of the local are, as different people will construct their local spaces in different ways, and these visions may conflict. One also cannot assume that all relevant local viewpoints will be organised and represented or able to develop their arguments so that they can be well understood by organisers. Individuals or groups may, for instance, not be able to express arguments in relation to the technical language often used by consultants, or they may not be used to putting their fear into words. Despite a relatively straightforward process compared with other levels, then, there is no ready-made locus of "community" waiting to be included (Blake 1999), as will also be discussed in the next chapter.

Thus, already on the local level, participation includes problems such as the need to coordinate across a vast number of organisations and individuals, such as the private sector and interested stakeholders (Sisk et al. 2001). Including the very different groups often targeted in participation also entails communication problems, which are marked already in rather immediate cases. These cases may include, e.g., the discussion of storage facilities, that road development, or a new industrial plant – which may be seen as the most concrete, tangible level of issue to which participation can be referred, for which stakeholders might then be assumed to be the most motivated to participate in technical discussion. However, for instance, in research on the public understanding of radiation hazards, Wynne discovered that people living close to Sellafield knew little about basic radioactive processes, such as the different properties of radiation. More significantly, however, they did not feel a need to know. This was something that Wynne eventually came to see as an altogether functional response. As the scientific understanding had already been expressed by various scientific experts in relation to the design of the plant and its operating procedures, workers simply learned the organisational procedures, not the science, and placed their confidence in the institution. Wynne noted that this is something that scientists also do in areas in which they themselves are not experts, to an extent that is seldom correctly recognised (Wynne 1991). For instance, most of us do not learn exactly how cars or refrigerators work in order to use them, but we simply assume that they should work. A problem in this regard is that a scientist or expert typically expresses issues in their own language and on their own level of concretion: what they need to know. However, the public may not need to know exactly how the car (or plant) is developed; what they need to know is what the consequences and security hazards are. As a result, if the expert-level or even expert-relevant knowledge is the technical level to which discussion is targeted, one can see that even concrete cases may result in high technical demands and a difficulty to motivate participants and attain participation.

At a national level, then, the problems salient for the local level are extended. Communication here takes place largely outside the common context of shared residence or shared concerns that a local development may result in. Also, on the national scale, concerning participation regarding policy change, for instance, it is immensely difficult to select and bring together relevant actors. "The public" becomes an almost unlimited group, with the result that consultation is almost generally limited to powerful institutionalised interests and government. Nevertheless, national consultation has an advantage over international consultation in that it takes place in a setting for which the legislative and other regulatory norms are determined through the national system. On the national level, there is a predetermined framework and authority structure that governs action and may provide legitimacy and clear output to participation, in that it may, for instance, be intended to support national decision-making in a directly elected constituency (Nanz and Steffek 2004).

However, a further complicating factor at the national level is that consultations may not only be those that have developed at this level. In the European Union (EU) case, directives taken at the EU level are to be directly integrated in member states, and consultations or processes that result from them may thus not always be fully integrated in highly varying state systems (Hooghe and Marks 2001a, 2001b; Pettersson et al. 2017). Participation in relation to an EU focus on participation – for instance involving the Water Framework Directive – may in some cases conflict with national conceptions of governance in the area. For water, thus, it has been argued that EU requirements on who the stakeholders are in relation to catchment areas may conflict with which actors are required to act in relation to national legislation (e.g. Keskitalo and Pettersson 2012). In early implementation of the Water Framework Directive in Sweden, questions included, for instance, that if stakeholders in such processes suggest specific requirements, who should implement them, and how this should be funded (Keskitalo 2015). Thus, perhaps in relation to the variety of approaches and understandings that are encased at the national level, problems involving the integration of EU regulation with the national level have been shown in multiple cases (such as water, forest, and minerals; Moss 2004; Pettersson et al. 2017).

On the international level, then, governance is even more remote from citizens and from the specific contexts that decisions at this level may impact. Participation procedures may be more opaque and may be dominated by diplomats, bureaucrats, and functional specialists as well as interest groups, potentially even focusing on interest groups certified within specific processes. Although the foundational legal acts of international governance are often subject to national ratification processes, everyday norms and standards are also often negotiated by non-elected experts and government officials. This typically occurs without the inclusion of mandated public representatives and interest groups in decision-making processes. International organisations, similarly, do not necessarily ensure adequate information to or policy choice discussion with interested citizens, as they may be far removed from specific contexts even at the national level. On the international level, "citizens" also become an immensely varied group, largely lacking shared context or given entry points from which to select representation in decision-making processes (Nanz and Steffek 2004).

It must thus be noted that processes on even higher levels that aim to provide guidance at the national and lower levels may indeed have very different effects in different cases. As decision-making today is often seen as undertaken within a context of multilevel governance, that is, including governmental administration at local, regional, national, and in relevant cases EU and international levels, as well as market actors and non-governmental organisations, understanding the different bodies impacting decision-making becomes crucial. As noted earlier, different countries have different regulatory and legal systems, cultures, and resources, including differing levels of decentralisation (Urwin and Jordan 2008). Large differences in regulative requirements and situations may also exist within states (particularly emphasised within federal states), which means that framings or problems that are forwarded on an international level may not necessarily align with those defined on lower levels or in any specific national or subnational case. In fact, it could also be the case that framings on one level are counterproductive to a desired development as it is perceived on a national or subnational level (this can be compared with the discussion on how different policy instruments cannot necessarily be transferred between cases; see Chapter 3).

While none of these considerations mean that there is not a value to participation – on the contrary – they do mean that participation cannot be assumed to result in either more democracy or fairness. Rather, it means that participation that is to make a difference must be designed based on, among other things, a knowledge of the case and the participants for whom it is intended, so as to support the values intended by participation. Participation must also be designed according to both process and output (how to undertake it and where what results will be used), so as to justify the expense of participants' effort and assure that processes support democratic or fairness aims of participation, within a situation marked by institutional dynamics and power perspectives (Wesselink et al. 2011; Kochskämper et al. 2016).

Concerns involving assumptions about stakeholder participation in relation to climate change: examples from different areas and bodies of literature

Climate change is an extremely wide field, and participation in relation to climate change may have been practised in very many different ways. However, given the issues we have already observed in relation to a linear model of scientific knowledge, there are at least cases in which the process and concerns regarding communication across professional, locational, and power differentials have been relatively sparsely problematised in the climate change field, particularly historically.

An example can be taken from early integrated assessment (IA) participation as a means of interaction with and including stakeholders. IA participation developed relatively early on, largely as a distinct practice in relation to the climate change field (Van de Kerkhof and Wieczorek 2003). Largely developed based on the concerns of the scientific community, IA participation has often been seen as focusing on direct scientist/non-scientist interaction, often in group discussions with a primacy to the scientific understanding (Van Asselt and Rijkens-Klomp 2002). The focus seems to have been largely more on lower levels (not national or international), in this respect reproducing the focus on lower levels that has been highlighted in relation to direct action (e.g. in the IPCC, as discussed in Chapter 2). In IA participation processes, participation was largely intended to provide input to existing model development. As a result, consultation was often undertaken centring on model use (Kasemir et al. 2003). One approach to attaining stakeholder interaction, in this case, was the focus group interview whereby scientists and stakeholders discussed a number of scientist determined topics, for instance with the visual aid of a computer model (Schneider 1997; Moss et al. 2001).

However, some authors note that while the development in this case might have been specific, it could be seen as related to processes used in broader "sustainability science" (e.g. Kasemir et al. 2003). As a specific volume on this highlighted, "sustainability science" often emphasises the role of direct participation in relation to science (Kasemir et al. 2003, eds; cf. Reed et al. 2016); we have also noted in this book that this is a focus in extensions to the linear model of scientific knowledge.

While participation in relation to climate change science cannot be taken as any singular field, then, it is relevant to discuss some experiences, caveats, and concerns that have been found in conceptions that relate to ideal-type or linear model of scientific knowledge conceptions and that are found in literature relating to IA or broader sustainability science or environmental participation.

Some of the implicit assumptions we have discussed regarding the use of participation can be seen in statements on the communication and interaction process as a procedural, rather than substantive, issue (as criticised by, for instance, Tansey et al. 2002). In some cases the assumption has been in line with one based on ideal-type or linear models discussed in previous chapters, and it focused on the notion that learning and discussion will take place as a result of coming together. For example, Ravetz, writing about scientist-stakeholder discussions over models, says: "It could well be that in the *open setting* of these focus groups, *free of any of the commitments and prejudices that affect policy debate among vested institutional interests* . . . questions could be framed and expressed all the more clearly, and with greater force" (Ravetz 2003: 63, my emphasis). Describing interaction as an open, unprejudiced process free of interests, Ravetz continues:

> [I]n the context of our research it was *possible for plain people to speak plain words* about their confusions and reservations about these scientific instruments. And so the *learning experience became universal*; they learned about the climate change problems in relation to urban lifestyles, and the experts learned about the different aspects of the usefulness of their tools in the general policy process.
>
> *(Ravetz 2003: 63, my emphasis)*

These types of statements can be seen as related to, for instance, more Habermasian assumptions regarding interest-free open discussion, unhindered by, for instance, technical language. Similar to Habermasian assumptions, one main focus of these interactions, in both this case and others, has been that mutual learning should ensue. The processes are explicitly aimed at educating scientists and stakeholders in each other's viewpoints in order to bring about institutional learning or, sometimes, with more focus on educating the stakeholders. Thus, for example, Hisschemöller et al. argue that "[t]he goal of a PIA [Participatory Integrated Assessment] exercise is to *widen policy-makers' scope*, or, to put it in jargon: to change their 'cognitive map'" (Hisschemöller et al. 2001: 62, my emphasis; cf. Gough et al. 2003). The emphasis here is thus on the idea that researchers and lay participants should "'learn' while 'participating'" (Gough et al. 2003: 45).[1]

Results of this kind of consultation, however, indicate some of the degree to which scientists and stakeholders entered the discussion with differing assumptions concerning what easily understandable communication tools and concepts are. Lorenzoni observed that:

> [i]n the main, local stakeholders think and act on the basis of extremely short time horizons compared to the timescales over which scientists predict rates of climate change. Although most are able to consider future events over a medium-term time frame (i.e. up to the 2020s), they identified many short-term constraints upon their ability to prepare accordingly. Chief among these are financial and budgetary requirements.
>
> *(Lorenzoni et al. 2000: 151)*

On a related note, Van de Kerkhof and Wieczorek noted that in processes organised by science, common complaints by the organising agencies are that involved stakeholders want to defend their own interests rather than focus on the "common good" (often seen from the scientist's perspective) and that stakeholders have an overly limited level of scientific knowledge to take part in discussions (Van de Kerkhof and Wieczorek 2003). Authors also note that participants' reactions to consultation with computer models illustrated problems with the methodology: Their "spontaneous reactions ranged from deep disappointment to active interest and from active rejection up to deep emotional concern about the information provided by the computer models" (Schüle et al. 1998: 3, quoted in Dahinden et al. 2003: 113–114). Dahinden et al. note that in their case "most participants felt that the computer models were less instrumental for the exploration of policy options" and "not sufficiently user-friendly and transparent" (Dahinden et al. 2003: 253). Similarly, "few of them [the participants] thought that they themselves would prefer to use the currently available models in order to get acquainted with climate change issues" (Dahinden et al. 2003: 119).

This could be seen as an indication that learning may thus, in such cases and in relation to what has been discussed in previous chapters, be limited by what is regarded as insufficient or not well-enough targeted consultation, within a limited conception of how social reality forms the possibilities for the types of "ideal" interaction potentially intended. In accordance with the priority attributed to science priorities in the process, these problems in stakeholder consultation could be seen as related to an assumption regarding interaction more in line with those in the linear model of scientific knowledge, or a more Habermasian understanding of ideal communication attempted in real-life situations.

There was already an awareness of these problems in the field of, for instance, early IA participation. For instance, in 2003, Van de Kerkhof and Wieczorek presented a number of factors that they regarded as preventing stakeholder interaction from working. Factors include an irrationality of stakeholders, whereby they are seen as manipulable and "rational" only to a limited extent. They are also seen as selfish, meaning that they are "likely to defend their own short-term interests and to 'free ride' on collective goods" rather than act for the common good. They are also seen as often having an insufficient level of scientific knowledge to participate meaningfully. It is also noted that the low level of scientific knowledge among stakeholders is worsened by their often prioritising scientific knowledge rather low or mistrusting scientific experts. Conflicts often emerge between different groups in consultation, as they see to their own interests (Van de Kerkhof and Wieczorek 2003: 7–8). As a result, Van der Kerkhof and Wieczorek note (as also noted here), more participation is not intrinsically more democratic; it may instead reinforce the interests of the already powerful, who can access relevant information and decision-makers. Finally, as only a selection of stakeholders can be involved, there are also questions of representativeness (Van de Kerkhof and Wieczorek 2003: 7–8).

These problems, as highlighted by Van de Kerkhof and Wieczorek, thus seem to derive from the fact that the real world does not cohere with the assumptions placed on it: Stakeholder interaction does not "work" because of how stakeholders work. Nevertheless, stakeholder participation even now may not have fully taken these lessons on board (as further discussed later, e.g. Plummer et al. 2012).

While the examples here are largely taken from early IA approaches, and some examples of the focus on direct participation have been illustrated in Chapter 2 with regard to IPCC framings, it should also be noted that these tensions in the possible ways in which participation and stakeholder interaction are seen may to different degrees manifest in various bodies of literature related to environmental change.

As a further example, the focus on direct communication to support change or learning has also been expressed in adaptive management literature. In terms we may expect by now, in a systematic review of adaptive co-management literature Plummer et al. note that broad adaptive management or adaptive co-management approaches have been seen to place great focus on fostering collaboration as a means to bring about positive change and support adaptive capacity (Plummer et al. 2012). The authors note that "combining the learning aspect of adaptive management with the linking function of collaboration, was the most frequently cited purpose" found in their review (Plummer et al. 2012, section 4, para. 3).

However, the review by Plummer et al. also illustrated the limited definition of interests of stakeholder groups and the limited rigor in the assessment of participation that could be found in the literature at large (Plummer et al. 2012). Similarly, in a critical review of participation in adaptive management literature, Stringer et al. (2006) identified a number of issues around participation, similar to those discussed earlier. The authors noted both the normative aim of this literature and the uncertainty of process or mechanisms for bringing these normative goals into being. They write:

> The adaptive management paradigm treats knowledge about ecosystems as both uncertain and pluralistic. It recognizes that, to create more sustainable management strategies, stakeholders must forge new relationships to enhance multi-directional information flows, learn from each other, and together develop flexible ways of managing their environments. Although there is general agreement on the importance of citizen involvement within the adaptive process, it is unclear how best this should happen, and what form it should take. . . . [U]nderstanding what "participation" means specifically within the adaptive management cycle has been largely neglected.
>
> *(Stringer et al. 2006, section 1)*

These types of statements could be taken to suggest that problematisation of many of the issues reviewed here, such as stakeholder definition, issues of power in communication, and process and output of participation, may have been limited – potentially as participation has been a requirement and add-on in natural science studies in line with assumptions in the linear model of scientific knowledge. Participation as venue may thus not have been sufficiently problematised in a wider context of what knowledge is to be attained or alternative means to develop this. In this book, the argument has been that that the type of knowledge to be attained, regarding the implementation of scientific knowledge in a social setting, can in fact not be sufficiently developed through direct interaction; it requires institutional analysis of "stakeholders" in their setting, including an understanding of organisational and governance forms that may impact agency.

Yet another example of this problem can be taken from the broad literature on climate change adaptation, which we have seen in Chapter 2 developed based in impact studies, for instance. Many somewhat procedural assumptions can be seen as related to this basis. For instance, a relevant general assumption in much adaptation literature has been that win-win solutions should be identified and that at least maladaptation (adaptation that causes negative effects among other actors than those who undertake the adaptation, i.e. that transfers vulnerability) should be avoided. However, suggestions have often been made that the correct developments be mainstreamed, but they have not necessarily analysed the preconditions for how (e.g. Owusu-Daaku 2018). Many authors have noted that, to include the relevant scope of social actors in adaptation strategies and frame information on climate change in relation to specific needs, it is not sufficient to simply develop communication between different groups or even leadership (for instance, science policy) (Grothmann 2011; Agrawal 2010; Smith et al. 2009). Even relatively recently, Eriksen et al. (2015) highlighted the political nature of climate change adaptation, noting that the literature so far had only seldom gone beyond observing inequality to understanding how "climate change is yet another context for struggles over how values and resources are governed and accessed across scales" (Eriksen et al.

2015: 527; cf. Owusu-Daaku 2018). Eriksen et al. also called for better analytical tools for understanding how power works as a productive force and suggested a focus on the need to understand power and politics; or as they saw it, authorities, knowledges, and subjectivities (Eriksen et al. 2015). By this they meant, among other things, how different rules, norms, and organisations interact – in what has been highlighted in this book as part of understanding institutions – and thus create struggles over authority as well as specific understandings or subjectivities.

Thus, as this book has shown, whether different levels may act on adaptation may depend on a number of factors, most of which relate not to climate but to broader institutional factors. Institutional frameworks and the role of power in them could be seen as influencing any deliberative approach to participation, stakeholder interaction, or interaction at large – making mainstreaming less an issue of direct implementation or communication than of starting from a social understanding to tailor implementation to the institutional dynamics within which it needs to be undertaken.

Does a search for "solutions" in fact obscure the problem?

Much of the literature cited in this chapter expresses an assumption that science will in some way find "solutions" (although not always in these terms) to problems. Direct interaction will result in collaboration, which will improve management. However, as we have seen throughout this book, it cannot be assumed that science will be able to directly express knowledge in a way that is then acted upon by stakeholders. Assuming this ignores the social reality of stakeholders and the concerns they need to take into account in order to act, or that are crucial for an issue to even get on the operational radar in order to be acted on.

Thus, there may exist a need to remain cautious regarding the extent to which normative aims – or a search for direct "solutions" focused on interaction – in fact obscure the problem: Is "stakeholder participation" in fact the best, or only, way to go? There is also reason to discuss whether the assumption that participation will result in learning has in fact been more of an assumption – related to, for instance, the disproven linear model of scientific knowledge – than something that is tested and developed in relation to multiple established frameworks and institutional study in the literature (cf. Plummer et al. 2012; Stringer et al. 2006). There are multiple studies and developed fields of literature that point towards ways in which individual rather than social learning may take place, as well as specific conditions and prerequisites that may be supportive of learning (in line with, for instance, the discussion on receptiveness and position noted here) (e.g. Reed et al. 2010; Benson et al. 2016; Vulturius and Gerger Swartling 2015). However, these qualifications and in-depth knowledge are not always taken into account in models that fundamentally rest on assumptions in line with the linear model of scientific knowledge: that knowledge will lead to change and, in the modified version, that knowledge communicated to stakeholders will lead to change (see Chapter 1).

In this volume, as well as in much literature, the answer to whether "stakeholder participation" is the best way to go has thus been a qualified no. In line with the focus here on questioning the assumptions in the linear model of scientific knowledge, the suggestion is that the focus on bottom-up stakeholder participation, knowledge, and learning is a framing that obscures the other relevant parts of the system, and in fact precludes an open problematisation of where the problems lie. In this volume, the complementary and overriding framing chosen has been that of institutions: that stakeholders cannot be understood apart from their institutional context, for which direct interaction alone is not likely to be able to provide sufficient understandings (Manicas 2006).

Thus, if, instead of starting from assumptions in line with the linear model of scientific knowledge, one started from an understanding of actual social interaction, the emphasis would be not on making communication rational but on ensuring that people, individually and otherwise, are able to clearly articulate their "irrationalities". These would not even be seen as "irrationalities", but rather rationalities as formed and defined by their everyday constraints and considerations. Participation or knowledge-gathering on such a basis should also not be assumed to directly be able to understand, access, and change people's social worlds but should perhaps take the more modest and exact approach of starting with understanding them and gathering knowledge about them – in order to revise the processes that participation is intended to inform, for instance to enable the design of instruments that relate to responding to the best scientific knowledge about scientific problems.

In such an understanding, then, participation – including any scientific participation – would be less about changing people's understandings than about gaining information about them. A view of stakeholder interaction as something starting from stakeholders and based on stakeholder concerns would thereby place the emphasis on much larger change, far beyond technical measures or procedural considerations to only a participation-focused model to take into account social considerations.[2] Notably, it would include the social analysis that is exempted from the linear model of scientific communication.

As a result, understanding the social world would not only rely on direct participation or direct communication between natural science and stakeholders. Instead, gathering information about "stakeholder worlds" would rest on established social science methodologies for gathering such information, for instance legal, planning, and policy studies and interviews, potentially on multiple levels and matched with analysis of other data, based on different disciplinary foci (cf. Irwin and Michael 2003; Manicas 2006). A crucial part of this would thus focus on understanding "stakeholders" in context: through broadly institutionally based analysis and within the context of the power, interests, and institutional issues that determine and situate them.

Understanding stakeholder participation within a context of power and institutions

There are numerous implications for an understanding of "stakeholders", "participation", communication, and the numerous related concepts of the way social life has been understood in this book. The chapter has illustrated that environmental considerations are seldom the overriding priority for stakeholders, and to understand why this is the case, we need to recognise what the requirements related to which stakeholders act are. Often, they lie in economic or market considerations over a short time, which determine whether they at all remain in their present role (Keskitalo 2008; Lorenzoni et al. 2000; Van den Bergh 2013). This is not necessarily an issue to be criticised per se, as it is not likely that it will be possible to change; rather, as forwarded in this book, it is an issue that reflects social life and can be made subject to an institutionally informed analysis as to, for instance, what incentives may be able to influence economic requirements, among other things (cf. Keskitalo 2008).

A broadly institutionally focused analysis, of the types discussed in this book and that focuses on an understanding of stakeholders in context, can thereby be seen as a necessity. Stakeholders – like scientists, or in fact any humans – are seldom themselves constantly aware of or able to explain the multiple constraints upon them from several levels of actors, policies, and practices of both formal and informal kinds. Thus, for instance, stakeholders will be impacted by national and supernational legislation and directives for implementation and regional and local policy, issues that pertain to sectorial responsibilities and to responsibilities in one's work and private roles. All of these may gain influence on both the formal and informal level, and with the sum of pressures perhaps less consciously conceptualised than simply adjusted to or responded to on demand. This is because, through processes of socialisation, people's interests can be shaped even without their realising it (e.g. Bachrach and Botwinick 1992, as also discussed in Chapter 3, this volume).

This means that the main "problems" of stakeholder interaction are only problems if one sees them from a rationalist perspective, which assumes that interaction would be conflict-free, "rational" according to scientists' interests rather than societal ones, and possible to undertake in direct interactions that place the primacy on a scientific focus (Philips 1995). Instead, it is crucial to understand perspectives that are relevant to the broad social context and what is often treated in terms of stakeholder participation, and that relate to distribution and social justice issues.

In this book, power has been defined according to theoretical frameworks based on real-life empirical analysis. The intention in applying these conceptualisations has been to cover the highly varying faces of power and, particularly, perspectives that might not be expected from the viewpoint of a linear model of scientific knowledge or a Habermasian perspective, which would either exempt understandings of power or see them only as negative or controlling. In this framing, the general understanding of power can be seen as being in line with Barnett and Duvall's (2005) definition of power as the production through social relations of effects on actors (cf. Eriksen et al. 2015). This type of understanding highlights that power thus does not only imply a direct, specific, or controlling relationship but also a distant, mediated, and productive one. Power relations are thereby seen as omnipresent but also as possible to conceive of in different ways. Thus:

> Instead of insisting that power work through an immediate, direct, and specific relationship, these conceptions allow for the possibility of power even if

> the connections are detached and mediated, or operate at a physical, temporal, or social distance. Scholars that locate power in the rules of institutions, whether formal or informal, frequently trace its operation to such indirect mechanisms. Those examining concrete institutions have shown how evolving rules and decision-making procedures can shape outcomes in ways that favor some groups over others; these effects can operate over time and at a distance, and often in ways that were not intended or anticipated by the architects of the institution.
>
> *(Barnett and Duvall 2005: 47–48)*[3]

Seeing power in this way means that if stakeholders are really to be seen as "stake holders" – persons whose interests one wants to understand – participation must generally be designed to allow them to express these interests and considerations within the context of the power relationships they are part of and that influence them.

However, the book at large has also highlighted that the understanding of these power relationships in general is largely the purview of institutional analysis: Given the role of institutions in real life, it is not relevant to design participation as a single unitary social contribution without placing it in the context of a larger institutional analysis and a considerable social science contribution to avoid the pitfalls of a linear model of scientific knowledge.

Seeing stakeholder participation and any requirement for stakeholders to be "rational" in this light would thus place an entirely different understanding not only on the possibilities for participation but also for identifying (even distant) stakeholders and being able to gain their input and for the assumption that one might even be able to directly understand each other's preferences and concerns. Instead of seeing stakeholders as directly identifiable, directly possible to communicate with regarding scientists' considerations, and directly able to identify and voice their considerations, an understanding of power and institutions as foundational to social life would thereby instead place requirements on understanding stakeholders' rationalities and how they are placed in a context of power and institutions. As noted in earlier chapters: What are the institutions that impact what stakeholders can do – the legislation, policy, organisations, and traditions that influence them? What are they able to do within this context? Where are they located, and are they – for instance, in light of discussions in the previous chapter – perhaps not only possible to identify but able to act within a local case? Can change be accomplished locally, or where are the steering institutions for the particular considerations one is targeting situated?

Thus, understanding something like the possibilities to act on climate change is not a question that is possible to elucidate only through direct stakeholder interaction. The "stakeholder perspective", in an understanding focused on real life, is not directly accessible through only direct interaction or enquiry to the stakeholder and cannot be seen as unproblematically accessible through direct "participation". The "stakeholder" may not be able to explain or rationalise their perspective, may have naturalised some understandings of the world rather than be able to express them in conversation, or may not be able to express them in a way that they are understood and seen as valid by others around the table (Irwin and Michael 2003; Manicas 2006). What is more, the individual stakeholder may not even have resources to make it to the decision-making table, for instance if it is not a local event but a national or international one. The problem formulation that those organising the interaction have developed may also not fit with what the stakeholder conceives of as the problem (as will be further discussed in the next chapter).

While this does not mean that "stakeholder participation" is impossible – far from it – it means an approach to stakeholder participation must be taken that is different from any that would derive from assumptions in the linear model of scientific knowledge or from a Habermasian perspective. It would mean that "stakeholder participation" may itself be a research project – and the fact is that the types of foci that, for instance, the natural sciences place on stakeholder participation are indeed often practised as research projects by the social sciences, in a far larger context, theoretically situated, and connected not only to participation as a concept. Understanding "stakeholders" and "participation" requires a problematising of the social at large, including through institutionally focused analysis and also an understanding of any direct interaction in the real-life context we have discussed here. It is not power-free; nor are "stakeholder considerations" possible to understand apart from also conceptualising the broader social reality and institutional context. This implies that understanding "stakeholders" or "participation" requires both research or review of research into the institutions themselves, and a conceptualisation of processes of participation, in relation to an understanding of problems with simplified assumptions on knowledge and learning, as discussed in the previous chapter. This makes it crucial to first understand how these processes are channelled and understood through institutions and institutional constraints, incentives, and processes (Agrawal 2010; Ivey et al. 2004; Smith et al. 2009). A stakeholder is never isolated from the social processes; rather, the situation of stakeholders is created and possible to change mainly institutionally, and all interaction is subject to institutional and power processes.

Key points

- Stakeholder participation must be understood within a context of power and institutions.
- Identifying and communicating with stakeholders are subject to all the complexities discussed, for instance, in previous chapters as well as here on power and rationality. This means that to understand social context it is not sufficient to in an unproblematised way "talk to people" without also analysing the situations into which they are placed.
- Rationalities are formed by people's context: People can largely be seen to do and preference what is rational to them given their circumstances. It is thus unproductive to accuse people of limited rationality, as the real concern for

research should be to understand these rationalities ("stake holders") and the potential incentives that could then be developed to promote other ways of acting.

Study questions

- What is a stakeholder, and how can you identify who a stakeholder is? What are the contexts, potential institutions, and scales you need to take into account in order to understand stakeholders' social contexts?
- What is rationality?
- What are the ways in which direct participation could be added to in order to understand stakeholders' contexts?

Notes

1 For example, in the ULYSSES project, "[f]rom discussions at project meetings it became clear that most of the partners understood the idea of participation mainly as a way to democratize science and to empower citizens, rather than just as a means to improve the quality of IA. Climate change decisions imply that citizens must either accept the undesired consequences, or they must rearrange their lifestyles accepting substitutes for their usual habits. It was argued that people and communities have a right to participate in decisions that affect their lives, property, and things they value. The idea of IA focus groups was advocated as a way to involve citizens more directly in the formulation of climate policy" (Van Asselt and Rotmans 2003: 222). Notable here is the focus on individual-level action in response to improved knowledge, without similar attention to governance levels or, for instance, an institutional analysis to support incentivising change.

2 In early IA work, for instance, Van de Kerkhof and Wieczorek suggested, as a response to the problems of participation, that it is sufficient to improve the communication process through primarily procedural considerations – to achieve suitable group composition, commitment of participants, fairness of process, input of scientific information, and deliberation in dialogue, "using a transparent process for balancing pros and cons" (Van de Kerkhof and Wieczorek 2003: 12). On this basis, one might note that the idea of improving procedural considerations, such as deliberation in dialogue and commitment of participants, can to some extent be compared with the procedural constraints raised in the previous chapter as necessary in order to bring real-life communication closer to the ideal. In that chapter, it was also clarified that there are major concerns regarding the idea that stakeholders could be made to adhere to ideal conditions, as well as the notion that the result of the process would then be usable in a world not marked by these conditions.

3 Although Barnett and Duvall (2005) define components of power as well as its institutional features differently than has been done here, the perspective here is not intended to differentiate itself from this type of broad conclusion.

Additional readings

Eriksen, S. H., Nightingale, A. J. and Eakin, H. (2015) Reframing adaptation: The political nature of climate change adaptation. *Global Environmental Change* 35: 523–533.

7

UNDERSTANDING ENVIRONMENT, SOCIETY, AND SCALE

Why outcomes of the same types of measures are not the same everywhere, and the local level is not only local

Introduction

The ways in which climate change and adaptation literature has largely focused on direct "bottom-up" interaction, with a focus on "community" or the local level, was discussed in Chapter 2. The chapter problematised that this type of conception largely exempts how interaction is situated within the context of institutional features on higher levels. Chapter 4 illustrated this with a focus particularly on the sectoral and national contexts, showing that what could be done in a sector like the forestry sector – and thus also among individual forest owners – needed to be seen in relation to processes on larger scales, including the national level. Chapters 5 and 6 then further specified the role of such real-life processes with regard to the impacts on knowledge and learning, and the role of stakeholder participation, as well as the challenges to participation on different levels.

Taken together, these chapters should thereby have illustrated that the individual, stakeholder, or local level cannot be seen in isolation, and that direct interaction is not sufficient for analysing or conceiving of the individual, the local level, or "community" within a social context.

This chapter proceeds within this type of framing. In light of the multilevel conception forwarded throughout the book – that any study case needs to be understood in its national context as well as any others – this chapter aims to problematise particularly the focus on "community" and the idea that particularly the local level should be the best suited to an understanding of environmental issues. Problematising the local is particularly relevant to an understanding of adaptation, as literature has focused to a great extent on the notion that vulnerability as well as adaptive capacity will differ between areas, and that thereby "adaptation has to be local" (cf. Nalau et al. 2015). This understanding has also permeated IPCC assessment reports and broader environmental and climate change work and has perhaps

DOI: 10.4324/9781003043867-7

led to an overemphasis on the local nature of cases. Given that we have discussed throughout the book that the way science has developed has largely been tradition-bound – in what sciences are emphasised where and what framings they provide – it might not come as a surprise to the reader that the focus on the local can also be seen as related to specific historical assumptions.

Following this, the present chapter will go a step further in the understanding of how the categories that have been allowed to steer scientific inquiry are in fact based in highly specific assumptions. The chapter will highlight that the focus on the "small scale" or "local" can be linked not only to the added direct stakeholder interaction angle in the linear model of scientific knowledge but also to historical assumptions, on areas as broad as "society" and "nature". In particular, it will be argued that the historical division made between society and nature – whereby small-scale human settlements have been linked more to nature as large-scale settlements were instead definitionally seen as "society" – continues to play a role even today. These conceptions have led to an understanding of "community" that is marked by the idea of a distinction between human and environment, and that has been strongly criticised but still remains a part of our assumptions – in science as well as elsewhere. This is the case even though today, in the Anthropocene with humans being the strongest influence on change on the planet, we can hardly separate "nature" and "society" in this way or assume that only small-scale settlements are to be seen in relation to nature. The chapter will thus discuss how this focus on the local and small scale has manifested in different bodies of literature. Examples include how it has been pronounced not only in adaptation research but in socioecological or resilience research at large (e.g. Ernstson et al. 2010). Other examples include anthropological research on the community as the unit of analysis, with "community vulnerability" being a prominent theme from early on in adaptation literature (e.g. Ford et al. 2016; Hovelsrud and Smit 2010, eds; cf. Mimura et al. 2014).

Thus, the chapter proceeds from a discussion of the division between "nature" and "society" to one of "community studies". This is followed by a discussion of the issue of levels or scale as they are comprehended differently in much natural science-based literature compared with social science literature.

In line with the general focus in this book, the chapter emphasises that there is a way beyond a focus on only the very small scale or on overarching global scenarios (as discussed in Chapter 2); this would be to highlight how local cases are necessarily governed and influenced through higher levels as well, using an understanding of institutions and the multiple influences on decision-making to highlight features through which local cases can be understood in broader, theoretically based, context. As a final illustration, these issues are exemplified based on the forest case.

Understanding environment, community, and the local – in context

As mentioned in earlier chapters, framings that separate, for instance, studies of nature from the social – and through this, follow up an assumed separation between "nature" and "society" – are pervasive. The fact that studies of the environment have to such a high extent been conceived of as based in the natural sciences has historical reasons related to the development and focus of the disciplines (Noble et al. 2014; Noble 2019). These types of divisions in what studies of "society" or "nature" are seen to focus on can even be traced back to times when "the environment" could perhaps more easily be seen as conceived of apart from "human systems" (e.g. Borsboom 1988; Ward 1977).

These types of framings continue to form conceptions today. Examples of this include how IPCC assessment reports highlight small-scale traditional local communities, whereby the focus can be placed on direct interaction towards learning (as discussed in Chapter 2). An additional feature that has been discussed in this regard, and that can particularly be understood in relation to the emphasis on knowledge in the linear model of scientific knowledge, has been the role of traditional knowledge: the idea that small-scale communities and particularly indigenous ones carry traditional knowledge, which then needs to be taken into account in a role supplementary to that of scientific knowledge (e.g. ACIA 2005).

Here, the intention is not to dispute that specific communities, groups, or individuals carry specific knowledge – it has been a focus throughout this book that communities, groups, and individuals need to be understood in their specific context. This context necessarily includes the types of institutions they are part of. The intention is instead to dispute the assumed unitary nature of these institutions: that it would be only traditional knowledge that is important, and that it would mainly be indigenous or local communities that carry it, as well as that knowledge of a kind that is relevant for science to understand would mainly reside at a local level. Instead, an understanding focused on institutions would argue that everyone not only carries knowledge of different types but is also situated in different ways that need to be understood. It would also highlight that local cases can seldom be understood only though local study, but they must rather be understood as impacted by legislation and policy at multiple levels.

The focus mainly on the small-scale, traditional, community, local, and indigenous as particularly related to the environment can thereby not be seen as a given, but as a result of specific conceptions. As noted regarding other topics earlier in this book, where conceptions come from or what they exempt is not merely a theoretical or abstract matter; specific conceptions carry specific framings – like the linear model of scientific knowledge does – and failing to recognise these is done at one's own peril. The risk is that conceptions are forwarded that in fact reflect assumptions about reality instead of studied, analysed, and empirically proven parts of reality.

Tracing the origins of society-nature distinctions and the focus on small-scale traditional communities

Given that the interrelation between "society" and "nature" covers a very wide scope, the different traditions that become relevant can of course be described in highly different ways. However, the perhaps most prominent example in literature that historically distinguishes nature from society (or "culture" from the "environment" or "nature") had a great deal to do with colonialism and industrialisation. This distinction harks back to a delineation that was created when the dominant powers invaded "new" lands, conceiving of this as pushing "the frontier" ahead of them (e.g. Turner 1921). In this conceptualisation, the "frontier" was the meeting place between "society" (that of the invaders) and "nature" (that of the new lands, where "indigenous" people lived).

The difference might thus be seen as one of conquest more than anything else: While the invaders could of course be seen as indigenous in their own homelands, they saw themselves as an advanced society compared to people they defined as different from themselves and as a part of lands to be conquered (Glacken 1976; cf. Macnaghten and Urry 1998).

What was done was thereby a work of separation, saying that civilisation or society had the right to utilise nature – including the people who were seen as being a part of it. As a result, in this regard nature was conceived of as pure, clean, and undisturbed "wilderness", and the people present in it were conceptualised as living in small-scale "communities" rather than advanced societies. The concepts against which these areas and groups were defined, notably society, culture, or civilisation, were on the contrary conceived of as outside and differentiated from nature or environment (e.g. Cronon 1995, 1996, 1987; Prout and Howitt 2009, Büscher et al. 2017). It is important that the difference in how "wilderness" was conceived of is thus definitional: Any larger-scale communities were seen as "society" instead of "nature", despite the fact that they would of course be situated in natural surroundings.

Archetypal examples of this sort of thought can be found in research historically. Turner's extremely influential work, in a book first published in 1893, popularised the idea of the frontier as the separation between man and environment (Turner 1921). Around the same period, the influential sociologist Tönnies published the book *Gemeinschaft und Gesellschaft* (first published in 1887, translated into English as *Community and Association*) (Tönnies 1955). These types of frontier thinking (Turner 1921), and related wilderness thinking (e.g. Cronon 1996) as well as thinking that separated large-scale societies from communities assumed to be more primordial (e.g. Tönnies 1955), popularised and provided a focus for the idea that nature and society should be conceived of as being apart from each other.

In some of the language employed in this book, this can be seen as a "framing" of nature and society that results in both assumptions regarding what small-scale communities should be like (living off the land rather than industrialised; small-scale; and not connected to larger society or multilevel systems) and what nature should be like (to be conceived of as being apart from human influence or human systems) (Ward 1977; White 1991; Redclift 2007).

Criticism of essentialist understandings of characteristics

As has been shown, the idea of nature was thus not only connected to environment but to people. People who were seen as living "in the environment" were conceptualised as less advanced and more primitive, but also conflict-free and traditional, while industrialised society was instead described – notably, by industrialised societies themselves – as symbolising "progress" and "advancement" (e.g. Kashima et al. 2011).[1]

This image – like the undead linear model of scientific knowledge – continues to mark current research and criticism. In current work as well, it is often identified – and criticised – that local and "indigenous" communities have broadly been assumed to be closer to nature and more conflict-free (e.g. Philip et al. 2013; Carter and Hollinsworth 2009; Sherval 2009). Similarly, particularly local and indigenous small-scale communities have been assumed to be more "primordial" and "traditional", and seen more as "communities" than what has been assumed in regard to industrialised or more large-scale localities (Glackin 2015; Goodwin 2012).

These assumptions are criticised in large bodies of literature, which note that these are essentialist framings that present highly varying characteristics as unitary and unchanging. For instance, Sherval notes that this is done to the extent that localities that are conceptualised in this way may even find it difficult to "develop" economically as they are seen as escapes from stressful city life and are to remain traditional (Sherval 2009). In this, they are sometimes situated to continue being something they perhaps never were. At the same time, the linkage to the environment – food and life support – of the city dweller as well as society at large is obscured and denied, as it is mainly the small-scale communities that are seen as linked to nature (e.g. Scott 2008).

Similarly, much literature has explicitly criticised the fact that smaller-scale communities have often been assumed to be more homogenous, close to nature, or even conflict-free (e.g. Philip et al. 2013; Carter and Hollinsworth 2009; Sherval 2009; Scott 2008). This idea that local community would in some way be more homogenous can be seen as being related to (or even potentially having come to influence) assumptions concerning participation in relation to the linear model of scientific knowledge: If there were no conflicts in society and everybody thought alike, it would be possible to gather information on society at large through direct participation. However, as noted earlier in this volume, the fact is that conflict may be as prevalent in smaller communities as in larger ones, and there may also be a multitude of different understandings and interests concerning issues, to the extent that attempting to gain "community participation" by only a few persons, for instance in a development, may even underscore conflict and support those who are locally powerful, to the detriment of other local groups (e.g. Holmes and Scoones 2000).

There are also important multilevel connections that are left out by a focus on the local level only. As briefly described in Chapter 2, it has generally been noted that even small-scale communities are today part of a much larger and even global context, where they may both be part of the production of goods sold on a world market and be reliant on world market produce. Even smaller-scale "traditional" communities can thus not be assumed to rely only on local production or subsistence, but may today also be relevant to conceive of in relation to a market framework. In addition, even small-scale communities are also often well integrated in national and subnational contexts (Keskitalo and Nuttall 2015). This means that decision-making even on, for instance, conservation or use of resources cannot be undertaken locally, but may rather be governed by national legislation and policy as well as by larger-scale actors. (The forestry case outlined in Chapter 4 can be seen as a case in point.)

These ideas of a separation between larger-scale societies and communities close to small-scale nature, whereby "communities" are then perhaps those targeted for participation in relation to issues on the environment, can thus be seen as something that does not relate to fact so much as to a conceptualisation of the world for historical purposes of conquest. Today, during what has largely been described as the Anthropocene – where human influence is so pervasive that it is changing natural systems across the planet – it is clear that humans everywhere influence and have a relation to nature (e.g. Palsson et al. 2013); the large variations in types of direct or indirect relations to nature (for instance through use or resources) may be crucial to understand rather than ignore if the aim is to actually conceive of human-nature, or social-environmental, relations.

The continued role of nature-society distinctions and community foci in research

Today the idea that social and environmental systems would not be connected is generally formally confronted in research – almost to the extent that it would be surprising in an environmental research context to state anything else (as visible, for instance, in discussions on socioecological systems, e.g. Berkes et al. 2008, eds).

Nevertheless, the division between humans and nature that was enforced through the conceptualisations discussed here remains an assumption in much literature (cf. Jørgensen 2015). Some examples of this include the fact that studies of the environment continue to be conceived of as based in the natural sciences, and the thorough roles of social systems as well as interlinkages between human and environmental systems have been relatively sparsely integrated (as authors on the Anthropocene such as Palsson et al. 2013: note; cf. Henderson 1993). The idea that social and environmental systems are interconnected has often been seen as a basis for assuming that analytical frameworks should encompass the study of both (such as in resilience thinking; cf. Olsson et al. 2015); however, assuming in this way that studies have to be undertaken through the same framework to give them the same weight may instead obscure the well-developed frameworks specifically on social study, such as those highlighted here, that describe social systems in the terms developed for social study (e.g. Brulle and Dunlap 2015). And similarly, a focus on wilderness as natural, unspoiled, and distinguished from – rather than part of – human use, also remains to this day, emphasised not least in recent rewilding literature (e.g. Nelson and Callicott 2008; Jørgensen 2015). This type of literature often highlights the importance of removing management from nature, rather than understanding the management traditions that exist and the impacts that removing them would have (Jørgensen 2015).

Examples of how these continued distinctions have played out in climate change and broader socioecological research are discussed in the following text.

- ***Community studies in adaptation and the Arctic Climate Impact Assessment***

The focus on the small-scale, community level may have been strengthened in relation to the focus in the linear model of scientific knowledge on direct stakeholder interaction. As a result, the "community" level might have been one type of study preferred in relation to climate change. This is the type of framing we identified in Chapter 2 as a problem in the IPCC reports: A large-scale (there, scenario- and modelling-oriented) understanding of natural science was added to mainly through a bottom-up focus on "communities" and community participation (cf. Mimura et al. 2014; Noble et al. 2014; Noble 2019). These parallel trends in study may, in fact, have contributed to adaptation having been relatively sparsely integrated in the broader body of climate literature (e.g. Noble 2019). At the same time, potentially in relation to the social focus being placed on the community level, the institutional, organisation-focused, large-scale components have often been less of a focus compared with an assumption that it is the small-scale, local, and close-to-nature cases that should be understood in order to understand human-environment relations (as noted in Chapter 2).[2]

A case in point of how this smaller-scale community focus has gained impact can be found in one of the early, large, climate change impact assessments, the *Arctic Climate Impact Assessment* (ACIA, published in 2005). Here, as the name indicates, a major focus was on climate impact studies. The human focus, however, was largely placed on smaller-scale and largely indigenous "close to nature" communities, as well as impacts on their traditional practices and how they might adapt (ACIA 2005) (see Text Box 7.1).

TEXT BOX 7.1 AN EXAMPLE AND CRITICISM OF THE SOCIAL FRAMING IN THE ARCTIC CLIMATE IMPACT ASSESSMENT

The Arctic Climate Impact Assessment (ACIA) study constitutes a major impact assessment, consisting of a large main document and several supplementary ones. Similar to discussions on other climate change assessments earlier in the document, the concern here is with neither the natural science nor the accuracy of the presented social focus with regard to the information it presents. The concern is instead with the number of important features that are left out of the social framing. In the short ACIA Highlights document the main description with a social focus reads as follows:

> "Climate Impacts on Indigenous People Many Indigenous Peoples depend on hunting polar bear, walrus, seals, and caribou, herding reindeer, fishing, and gathering, not only for food and to support the local economy, but also as the basis for cultural and social identity. Changes in species' ranges and availability, access to these species, a perceived reduction in weather predictability, and travel safety in changing ice and weather conditions present serious challenges to human health and food security, and possibly even survival of some cultures. For Inuit, for example, warming is likely to disrupt or even destroy their hunting and food-sharing culture as reduced sea ice causes the animals on which they depend to decline, become less accessible, and possibly become extinct."
>
> (ACIA 2004: 7)

Contrary to this type of main framing, it can be noted that the Swedish forest cases described in this book include areas that would fall under the geographical area included in the ACIA. As will be highlighted further on in this chapter, the Saami indigenous group in Sweden, also mentioned in the report, is highly technologically as well as market-integrated, as well as integrated with wider society. For instance, reindeer may be moved with trailer trucks or observed using helicopters or snowmobiles, herders typically maintain both summer and winter residences in order to be able to follow the herd, and reindeer meat is regularly sold on the market (Keskitalo 2008). There are no designations of particular indigenous areas (similar to North American reservations) and thus there can be no designation of Saami-only "communities" that are constituted by local, clearly delineated habitations. Identities and ethnicities are also typically rather mixed in northernmost Sweden, and people may identify as local, Torne Valley Finn, Saami, Swedish, or any mix of these or others (Keskitalo 2004). As noted in Chapter 4, reindeer husbandry is largely practised on the same lands as forestry is, and while this results in large sectoral conflicts between reindeer husbandry and forestry, reindeer herders may also work in forestry and own forest lands (with the extent of conflict varying down to an individual level).

This type of broader analysis may be particularly important in the case of adaptation, as this seldom takes place only in relation to climate change, as social actors in fact need to "adapt" and respond to multiple stresses all the time (e.g. O'Brien and Leichenko 2000; Næss et al. 2005). Adaptation to stresses will thus depend not only on climate impacts but also on the opportunities and requirements afforded under not least economic and social considerations – which one may understand very differently depending on whether one mainly focuses on traditional, indigenous, and local features or also takes into account the broader system within which decision-making is nested.

While criticising this is not to say that there are no small-scale traditional communities in the geographical area chosen for the assessment, it can be noted that the larger-scale, more institutionally or organisationally focused aspects were not those placed in focus. This is despite the fact that the role of larger-scale systems has been emphasised particularly for the northern European areas that would be relevant to this assessment (e.g. Keskitalo 2004; cf. Doel et al. 2014; Dybbroe et al. 2010).

While some of these problems may be particularly emphasised in an "Arctic" framing – which, in relation to a broader history of frontier thinking, carries numerous connotations of emptiness rather than of people, as discussed earlier (Keskitalo 2004) – they are, as we have seen, not exclusive to this geographic conception. Ideas about community, not least following the influence of Tönnies (1955), have also to some extent been forwarded in "community studies" on adaptation, which have often been anthropologically based and focused particularly on transformations in small communities (e.g. Olsson et al. 2004; Ford et al. 2016; Hovelsrud and Smit 2010 eds) (see also Text Box 2.4. in Chapter 2).

While this is not to say that contributions are not valuable, it does say that the focus of study has not equally covered all parts of the social system or all aspects that must be considered in order to understand adaptation in social systems. Thus, systematically highlighting certain issues – framing the issue as well as research on it – obscures other understandings that are relevant to adaptation as well as more broadly to an understanding of actual socioecological processes.

- ***Systems science conceptions of socioecological linkages and resilience***

These types of historically developed and disciplinary encased foci have also come to influence the development of attempts at integration. Systems science has been used as one application for conceiving of "systems" more broadly (e.g. Von Bertalanffy 1972 for a discussion of its history). However, and perhaps necessarily, given the historical contradictions and distances between different types of research, systems science also does not place all research on equal footing. The focus on socioecological systems (SES), which developed largely on this basis and gained emphasis prominently from the 1990s (e.g. Holling and Gunderson 2002; Berkes et al. 2008), is one example of this. Attempting to link social and ecological systems often proceeds from an assumption that the "system" (for instance in a local case) can be comprehended to the extent that limitations, barriers, and the like can be defined (e.g. Resilience Alliance 2007; Olsson et al. 2015).

However, the assumptions in resilience literature have often been criticised for focusing too much on being able to delineate a single system. They have also been seen as resting on assumptions regarding stability and self-organisation that are only typical of certain theoretical points of departure in social science, some of which are no longer widely applied (as Olsson et al. quote, even "dead as a dodo" [Barnes 1995: 37 quoted in Olsson et al. 2015]). Olsson et al. summarise:

> The most fundamental obstacle here, we argue, is the difference in how resilience theory and the social sciences understand society – in terms of social

> systems, social relations, and social change. In essence, resilience theory is implicitly based on an understanding of society that resembles [mainly historically applied] consensus theories in sociology, according to which shared norms and values are the foundation of a stable harmonious society in which social change is slow and orderly – and where, in analog, resilience thus becomes the equivalent of stability and harmony or the good norm.
>
> *(Olsson et al. 2015)*

This type of assumption on society, sketched by Olsson et al., can be seen as contradicting the institutionally based focus highlighted throughout this volume, in which power and interest play a crucial role and lock-in and change are related to, among other things, benefits, routines, and understandings involving different actors. In this regard, it is a problem that neither systems science nor much socio-ecological research strongly takes into account the influences of theoretical conception on what features of a case study may be highlighted or the methodological difficulties of attaining a system understanding of social systems given social conflict and different interests groups (Keskitalo et al. 2016; Cote and Nightingale 2012). It is also a problem that socio-ecological systems work has typically highlighted lower organisational levels (Ernstson et al. 2010) – a fact that may be related to the assumptions regarding where linkages to nature should exist, in relation to the conceptions of "nature" and "society" discussed previously (that is, without equally recognising the role of larger systems and multilevel contexts under globalisation).

Thus, in some cases, socio-ecological conceptualisations may express assumptions similar to those highlighted in Chapter 2 in this book: that the system should be assumed to be directly possible to observe and describe, with barriers and limitations subsequently identified and then potentially removed. Chapter 2 discusses the concern that this may obscure a recognition that barriers and limitations may be inherent in systems and also differ depending on interests or positions of various actors in the system.

The differences in conceptions of scale in natural sciences and social sciences cases

The assumption that the local or "community" level would be the most relevant for environmental issues, or be more possible to comprehend in relation to tradition, indigenousness, or related conceptions than other levels, must thus be challenged. It must be seen as part of a way of thinking that was developed for certain purposes and that is not relevant as a distinction per se in relation to conceiving of adaptation more generally in the Anthropocene, where the local is impacted by multiple other levels, which in fact all in some way relate to and impact resource – and environmental – use.

It has been noted that, while nature and society have historically been constructed as separate, for specific purposes, much research today highlights the need to understand linkages between social and ecological areas. However, the earlier analysis also suggested that the assumption that this would require integrative frameworks whereby the same theoretical understanding is applied to society as to nature is the result of a misconceptualisation. To understand the linkages between nature and society, and the way we inherently depend on nature, we do not necessarily have to apply natural sciences-derived conceptualisations to society. What this book has suggested in general is that scientifically voiced conceptions of the social that are developed outside the social sciences should be treated carefully: There already exist developed conceptualisations in various developments on theory that are empirically based and can highlight a linkage to nature in use, national- and lower-level frameworks, and the like (as illustrated in Chapter 4). In selecting frameworks that are not based in social study to describe social relations, one must thus consider what may be left out. Are the terms fully applicable? What are the assumptions inherent in different conceptualisations? Why is one theory, rather than another, chosen to explain or understand a phenomenon?

At the least, understanding linkages between nature and society does not mean that society has to be understood through similar frameworks. An example of this can be taken from the case of scale in natural and social sciences, which is again relevant to understanding the role of the local level and how to conceive of it in an empirical, realistic way.

Understanding the way scale is defined in social contrary to natural science studies

As we have seen throughout this book, understanding even the local level may require detailed study of relevant institutional contexts on different levels, potentially including the national, sectoral, or other regulative styles or logics relevant to the case. Thus, seen in the theoretical contexts discussed so far in the volume, even if local cases may vary on a wide scope, there are also things that can be known about them based on, for instance, a general knowledge of potential institutional dynamics. This means that, beyond the acknowledgement that adaptation must *to some extent* be local (Nalau et al. 2015) because there are differences between countries, regions, and localities, adaptation – or other environmental action – must also to some extent take place or exist on other levels in order to support local adaptation. There are things relevant to the local level that are governed from higher levels, such as the regional, sectoral, or national – issues that determine what can be changed and what not may even pertain to more overarching organisational logics (institutions such as the market or capitalism as a system) (e.g. Cashore 2002; Cashore et al. 2004; Marks and Hooghe 2004).

All this means that, in order to problematise the local scale, beyond understanding that the focus on "community" scale as most prominent may to some extent be based on historical assumptions that have been widely criticised, it is also important to understand that "scale" or "level" cannot be assumed to work the same way in trying to understand social systems as it does in trying to understand purely ecological systems. Whereas the natural sciences, for instance, may be able to delineate a landscape and focus study on this without per se taking in higher levels (Reed et al. 2016; Keskitalo et al. 2016), it is not possible to identify what social issues and institutions are relevant locally through merely local study (absent an understanding of broader society). This does not have to do with a notion that social and ecological systems would not be linked, but rather with the ways in which social systems at the local or lower levels are linked to systems at other levels as well. For instance, the national "scale" (a more general term) or "level" (a term more often applied to, for instance, governance levels) can thereby be seen as present locally as well, in how it may steer what can be done at the local level, for instance in the case of a local council or municipality, as noted earlier.

Natural and social scales may not "fit" with each other

While a problematisation of social and environmental assumptions may by now – through the discussion throughout this book – seem rather self-evident, it has thus seldom been recognised in discussions of scale that explicitly aim to bridge natural and social systems. Instead, despite the many difficulties we have seen so far in combining natural and social data, it has often been assumed that single scales such as the local can be chosen as a way to integrate or "fit" social and environmental organisation (Nalau et al. 2015; Ernstson et al. 2010). The idea here is that there should be a "fit" between ecological and social scale, so as to make problems manageable on the local level. This would mean, for instance, trying to organise local social systems so as to govern natural resources, in the way that "SES research has often argued that social systems should be organised in relation to ecological systems" (Keskitalo et al. 2016: 744; cf. Silver 2008). This type of assumption can also be seen, for instance, in adaptive co-management and landscape management literature, as well as in ideas about other types of environmental management initiatives such as model forests or biosphere reserves (e.g. Olsson et al. 2004; Sinclair and Smith 1999; Reed 2019; Reed et al. 2016).

However, in a comparison of different conceptions of scale across the natural and social sciences, Keskitalo et al. (2016) note that this type of "fit" to make problems manageable on the local level is not always possible. This is because the "development of potentially satisfactory fit of governance with ecological patterns and processes for one land user might impact negatively other land users with different resource management scopes" (Keskitalo et al. 2016: 743). What may be a fit for one type of user may thereby not be for another.

As a result, then, "it may not be possible to manage local conflicts by considering the local level only, nor by assuming that organisation will relate itself to ecological processes or be able to fit with these" (Keskitalo et al. 2016: 743). Instead, there is a risk that the assumption that resource governance should be local will in fact result in, in the worst case, mismanagement. The reason for this would be that the influence from higher levels and processes – such as those we have discussed in previous chapters – is not taken into account: If systems are treated as local even if they are not, solutions may be sought at a level different from where the problem conceptualisation or possible solutions may exist.

"Mismatches" must be assumed and researched

Keskitalo et al. (2016) discuss these types of problems of fit in terms of mismatches, noting that there are several types of mismatches that this type of thinking can result in.

One typical example is *spatial* mismatches, whereby a phenomenon at one level (such as the local one) may not fit with it at another. This might be one of the most common issues in relation to the discussions so far in this book; we can now use our earlier discussions of, for instance, framing to understand how the same types of problems may be defined and indeed treated differently at different levels. A problem may, for instance, be framed one way on the international level while it is framed and seen as a problem nationally or sub-nationally for other reasons. Problems that are encountered nationally or locally – for instance invasive species management, as exemplified in Chapter 2 – may also be framed or determined by other levels in terms of entirely different problems (such as free trade). This means that governing frameworks at higher levels may also influence local levels, to the extent that decisions cannot be made entirely at the local level.

The example also illustrates the risk of *functional* mismatches, in which the potential scope of solution does not fit with the process causing the problem. This is because, for instance, problems and the need for solutions may manifest at other levels than those from which the sector, for instance, is governed and the problem is created (Cumming et al. 2006; Guerrero et al. 2013; Keskitalo et al. 2016).

It could thereby be conceived that many of the issues that arise around trying to develop local governance in a simplified way that focuses particularly, or only, on the local level are in fact the result of ignoring the types of features that have been emphasised throughout this book. Social systems, as this book has aimed to show throughout, constitute different logics to those of ecological systems:

> In comparison with a focus . . . in ecological scale, thus, the pattern of governance or steering . . . differs widely in local, regional, national and international configurations between cases in or related to different countries, as numerous different regimes on different levels – such as international trade, the general broader institutions of the state, or specific regional or local configurations – influence any given resource use. . . . As a result, the socio-economic and political system must be seen as one where scale, i.e. the explanatory dimension, is not a given but as one where scale is constructed differently by different actors, with different features highlighted in various theoretical conceptions.
>
> *(Keskitalo et al. 2016: 744–745)*

As a result, any phenomenon related even to specific cases or areas for natural resource management will likely have somewhat different definitions at different levels: Rather than being an issue of fit – which suggests that it could be easily corrected – it is a matter of how these systems work and need to be understood. For instance, issues relevant to forest at the European Union level may cover a range of issues whereby, as we have seen earlier, international trade may also become important. These may then differ or partially overlap with how a state or other actors define issues relevant to forest or forestry (as discussed in Chapter 4). Different actors or interests may conceive of the "problem" differently, as highlighted in framing literature, for instance (see Text Box 5.1 in Chapter 5). As a result of these processes and the different ways in which problems at various levels may be defined and also prioritised and contextualised, the problem to be solved – or that is placed in focus at that specific case – may similarly be defined differently in various different cases. The local or community level will both be impacted by and be part of this variation in definitions, depending on the issues in the specific case.

Thus, rather than simply calling this complex problem a functional mismatch, as if it could be corrected by a "functional match", these definitional and interest differences will be inherent to multilevel systems (as well as, in fact, to smaller or even single-level systems; cf. Nagendra and Ostrom 2012).

Similarly, as historically developed power relationships often result in the maintenance of specific resource access rights, as we have seen earlier, resource access must also be seen as an inherently political issue rather than as one that can be corrected by "improved fit" only, as if it were an apolitical matter (Keskitalo et al. 2016). Instead, as emphasised throughout this book, it is important to understand how this politics works and how scale and decision-making are in fact constructed in these multilevel systems. Social systems must be recognised as being based in political, economic, and other interests and organisations developed over time, commonly encasing specific conceptions of environmental systems as well.

Scale is constructed depending on the case

There are thus concerns involved with attempting to define the local level only, or any one social level only, as the level at which a problem is "solved". One way to conceive of this situation in relation to institutional or power perspectives can be to see scale as constructed rather than given: One needs to understand the scale in relation to what other scales influence and thereby construct it.

As noted earlier, the ways definitions are developed at different levels are themselves often subject to the types of institutions, including interests and power structures, that have been discussed in previous chapters. This means that a crucial role in understanding these types of systems lies in institutional analysis not only of the local level or stakeholders but also of the ways different levels and institutions construct the local level. The idea that something is "constructed", then, is not intended here to refer to something abstract; rather, it is the highly practical way in which different interests may appeal and make themselves relevant to specific levels – if you wish, the construction of stakeholdership. But in this, it should also be considered that not all actors will limit themselves to action only on certain levels; active and multilevel roles are often taken by strongly invested interests, for instance in processes of jumping scale (Gupta 2008).

Such participation or construction of scale can thereby be seen in light of the discussions on stakeholder participation and power in earlier chapters. This means that actors – for instance organisations – may even attempt to construct or manage scale so as to retain or increase resource rights and "jump scale" to gain influence. Local actors may aim to influence local resource rights through participation in international work, the results of which they can then appeal to from local or national levels (Gupta 2008). Such dynamics further complicate issues of participation; as discussed in the previous chapter, "local" participation cannot be assumed to represent all local actors and interests. Rather, given the resource and organisational costs of participation, participants "jumping scale" may in some way represent specific interests acting to increase or defend resources in an area that is seen as particularly pressing or where the selected venue is accessible and may seem to offer particular promise (e.g. Pettersson et al. 2017). Also, not all actors that functionally influence the local level may see themselves as local – which may hold true even at the individual level (discussed later in this chapter).

For any community or local study, then, this would imply that it is not sufficient from a social perspective to focus only on the local level; instead, it needs to be understood how other levels impact this one and the actors in it, as well as how they may themselves influence other levels by jumping scale. In the case of strong interest groups, these may likely be present on both the local and other scales, and for instance draw on their substantial resources to represent at the local level as well. A local or community perspective can thereby also never be taken at face value as representing much larger local or group perspectives in a simplified way, but it must rather be seen as situated in the institutionalised power context that has been the focus of this book. A local stakeholder-based or bottom-up development may represent only certain interests or orientations and cannot per se be assumed to represent a community. As discussed in this and earlier chapters, local or smaller-scale levels or communities are not conflict-free, and nor can consensus be assumed. In addition, problems may have been created on other levels or may even be defined differently on other levels to the extent that local solutions cannot be developed.

The complexity inherent in scale as well as in conceptions of the "local" or "community" case in a real-life perspective is thereby considerable. Again, there are no shortcuts to describing this complexity without attention to actual research on the case, with the need to include the institutions that impact it and that it is part of. Thus, quite simply describing systems as multilevel governance ones or polycentric (e.g. Hooghe and Marks 2001; Nagendra and Ostrom 2001), or through a more general resilience or systems focus (e.g. Olsson et al. 2004), does not necessarily suffice. Polycentric, multilevel systems are subject to the same mechanisms as any institutional systems – and perhaps even more so, as they often exist over several levels where only stronger actors may be able to gain considerable influence, supporting for instance the role of economically more powerful interests at the cost of those lacking time, funding, or other resources for participation (see e.g. Pateman 1970; Nanz and Steffek 2004; Jentoft 2006). The fact that there are multiple aims within the system that often compete for resources and attention will also limit the possibilities for alignment on any one issue (cf. Kingdon 1995). In total, this means that:

> "fit" between scales will more often be a question of political compromise between parties involved, rather than of a one single good solution. In such cases, management of mis-match – including managing and clarifying different conceptions of a problem – may often be a more realistic point of departure than assuming that goodness of fit can be identified.
>
> *(Keskitalo et al. 2016: 750)*

Text Box 7.2 illustrates this need to take into account multiple social scales in resource management research, where a focus on landscape and specific local management or voluntary resource management bodies has often been prominent. The box illustrates that these types of specific management bodies must also be seen in the context of the broader decision-making system – particularly, they may also be seen as limited along the lines of general restrictions on voluntary steering mechanisms (as discussed in Chapter 3).

TEXT BOX 7.2 CONTEXTUALISING LOCAL PARTICIPATION IN RESOURCE MANAGEMENT RESEARCH

As discussed throughout this book, many larger-scale research programmes focused on the environment have come to include particularly a participative aspect (more so than a social study aspect), in line with an addition of democratic participation to assumptions in the linear model of scientific knowledge. Reed and Reed et al. (Reed 2019; Reed et al. 2016) forward this type of conception in a review of literature in relation to "landscape" conceptions, which are seen as one example of a "sustainability science" focus. They note that, for instance, many socioecological or landscape research cases (Reed et al. 2016) select for participation from local, sometimes informal decision-making or participative, fora with a specific focus on targeted resources (Reed 2019). One example of this is the originally Canadian-based and now internationalized Model Forest concept, which was conceived of as a venue for local collaboration around forest areas (e.g. Sinclair and Smith 1999). Other types of such often voluntary fora that have constituted a clear selection point, for instance for "sustainability science" initiatives, include local participative initiatives around sustainability (such as Agenda 21) and others (cf. Lafferty and Cohen 2001).

However, given what we have already noted on the multilevel and constructed nature of scale in relation to social systems, the local level is often not only local but also crowded with influences and requirements from multiple levels. The local level is typically already governed by municipal or local council structures, which may have formal decision-making power over land use and have to balance multiple demands (see Text Box 2.2. in Chapter 2). In addition, resource management is often governed by legislation, policy, and interest structures on multiple levels (as illustrated in Chapter 4), meaning that the specific area of governance falls under constructed scales of the type discussed in this chapter.

There is thus often already dense organisation present in the area for which specific voluntary fora are developed. As a result, it may be expected that voluntary organisations may per se be less embedded – and decisive – in the formal decision-making structure (e.g. Sinclair and Smith 1999). Along these lines, studies have for instance suggested that the degree to which these types of specific organisations work in different countries may depend largely on their relation to national governance systems (Elbakidze et al. 2010). While not detracting from the potential value of voluntary or delegated authority locally based organisations or networks, such fora thus need to be understood within a wider social analysis, including that of how they are able to forward their work within the wider organisation and within what decision-making frameworks and regulations.

Illustration: scale in the case of Swedish forest and forest owners

Much thinking on local cases in socioecological systems literature has assumed that the local case can be seen as a basis for management and conflict resolution (potentially akin to an assumption of a "community" as a fundamentally consensus-oriented structure). The local has also not always been seen as a nested structure that requires an understanding of other social levels, as discussed previously. Drawing on the case of forestry and forest use in Sweden, this section illustrates how the case of forest – like other cases – cannot necessarily be seen as formed, maintained, or influenced purely locally, and how the constructed nature of scale can be seen as manifested in practice.

As noted earlier, the long and specific histories of northern Europe may have implied a specific relationship to land as well as specific organisational styles: Fennoscandian areas stand out as "countries with a long tradition of [forest] associations . . . in contrast to countries with relatively new associations, including the USA" (Schraml 2005: 252). As an example, the significant role of forest owners' associations in Sweden and Finland, as well as the historical development and national nature of the forest industry, means that it is not sufficient to understand forest issues as local.[3] Contrary to seeing the forest as "empty" or "natural" land, it also is actually both full of people and also – in the Swedish case – heavily managed: actively planted, managed, and logged (e.g. Andersson and Keskitalo 2018).

This means not least that forestry production is part of a global system: Sweden is one of the world's largest exporters of timber, a competitive market in which local sawmills have closed, and production, including employment patterns in forestry, has shifted as the organisation and competition in forestry have changed over time (Keskitalo 2008). For instance, originally a Swedish industry, the now multinational SCA is only one of the industrial timber buyers that acts to a great extent in relation to competition on the international market. This means that what may seem to be local or national conflicts between production and protection interests may in fact be strongly influenced by the international situation.

The framework of global competition not only applies to multinational industry; it also impacts the room for manoeuvre of smaller local actors (for instance in the building industry, which draws on wood resources). The role of economic competition on an international market may even be seen as playing a role to the extent that any adaptation these actors undertake might be regarded to be in relation to economic incentives for business survival rather than to climate change (Keskitalo 2008; O'Brien and Leichenko 2000). In fact, in a 2008 study of forestry, reindeer husbandry, and fishing in northern Fennoscandia, it was seen that climate change mainly gained impact and resulted in adaptation as an influence on economic viability in all three sectors (Keskitalo 2008). This thus held true even for reindeer husbandry, which, while commonly assumed to be a traditional activity, is actually heavily technology- and capital-dependent: Reindeer migration may be undertaken, for instance, using trailer trucks, with herders moving between primary residences and potentially several second homes; herding and management may be undertaken using snowmobiles, all-terrain vehicles, and in demanding cases even helicopters; and reindeer may require, for instance, additional feeding during difficult periods. In the Fennoscandian cases, reindeer meat is also typically sold on the economic market and is thereby integrated in a market framework (Keskitalo 2008) (see also Text Box 7.1).

So local forest and forest use – although undoubtedly taking place at local sites – can thereby not be conceived of only in the local context. Neither, however, can local people be conceived of merely as local, in the way that they would be assumed to be possible to find on site. The way forest ownership is presently organised in Sweden means that it is not only those residing full-time in an area who are "local", in the meaning that they can claim to be stakeholders in the specific local area and local development. The well-institutionalised framework of forest owners' associations and industry means that it is possible today to own forest without knowing much about forestry, as services can be bought (Kronholm 2015; Erlandsson 2016). This also means that forest owners may not necessarily live on their property; in fact, many do not (Westin et al. 2017). However, this does not mean they may not be very attached to and concerned about their property or the area where it is situated (Nordlund and Westin 2011; Gunnarsdotter 2005). Properties may often be previous homesteads that have been inherited and are now second homes, as the present employment structure often necessitates finding employment in a town or city. There is also significant tradition around having a property outside the city – related to, among other things, specific practices like berry- or mushroom-picking and the like – which has been identified as applying to large parts of populations (Keskitalo et al. 2017b).

For these and other reasons, it has been assessed that about half of all Sweden's population have access to a second home (on a forest property or not) (Back and Marjavaara 2017; second-home ownership is also significant in Norway and Finland, for instance). Second homes thereby do not carry the same connotation of only involving a minority in this case, as it may in some other countries, but can rather be seen in relation to the historically dispersed population and land-use tradition. As there are no administrative means to know how much time is spent in "primary" compared to "secondary" homes, it is also possible that even individuals who list one location as "primary" with regard to where they are employed spend more time in their second home (Lidestav et al. 2017; Keskitalo et al. 2017b). Thus, the extent of movement between several locations, often including more rural and more urban ones, can be seen as significant in Sweden. To this should also be added the dispersed residential patterns of active reindeer husbandry, the relatively large hunting and fishing interests (Kagervall 2014), and the public forest use rights, e.g. for berry- and mushroom-picking, which mean that a forest can be embedded in interests far beyond the "local" area – and that even people who do not own property in an area may have significant interests in its use.

For these reasons, it is not necessarily given who is a "local" and should be seen as part of a local case. Neither is it given that local understandings would be particularly local, in that they might not reflect or be influenced by understandings elsewhere (Pettersson et al. 2017).

Resultantly, "stakeholders" may not be limited to those living on site, and strong emotions may be raised by second-home forest property owners who may have a strong investment in their properties and may also spend as much time in the area as any "local" (Bergstén and Keskitalo 2019).

In addition, the Swedish case can also demonstrate that, while local groups or individuals may indeed manifest behaviours or practices that make them seem close to nature, this does not mean they necessarily constitute any one united community or are in any way apart from higher levels. Forest ownership and the way forest owners, for instance, engage in berry- or mushroom-picking does not necessarily mean they are more "traditional" or "small-scale" in the meaning of "apart from larger organisation"; nor does this imply that they do not have other identities as well, or that they identify or could be seen as indigenous. In the Swedish case, given the historically developed role of, for instance, forest interests and the dispersed population, the role of what may be seen as "close to nature" practices is both relatively typical in regard to large populations as well as encased in and manifest in legislation. The role of forest ownership, but also potentially those of hunting and fishing or the right to public use, could be taken as cases in point (Kunnas et al. 2019; Pouta et al. 2006; Keskitalo et al. 2020). Hunting is a right connected to land ownership, meaning that many hunters are landowners in the same way they practise small-scale forestry and, for instance, may hunt traditionally for some time each year (von Essen et al. 2015; Ljung et al. 2014).

A linkage to nature thus does not entail a link only to "indigenous" or "local" identities (Keskitalo 2004), but it is rather constructed depending on the specific history and context of the area. As noted earlier, the assumption that indigenous, traditional, and community interests would be closer to nature must also be seen in relation to how these terms and categories have been constructed: in relation to conquest on a global scale rather than necessarily fully descriptive of actual local situations.

In total, interests that are invested in any local area may thus, in the Swedish case, rather than simply being local include a large variety, including state, industry, and member interest organisations with varying resources and funding, municipalities, small-scale owners in different sectors, and the public, residing either on- or off-site. All these actors are impacted in their use by multiplicities, different sectoral and general legislation, and regulation developed over time. Issues of concern to the local area may also be layered with more recently emerging regulation such as EU and international requirements, which may also define specific issues in different ways that may or may not "fit" with various local or national conceptions (as discussed in this chapter and in Chapter 4).

Conclusions

This chapter has shown that, contrary to common assumptions in socioecological systems research, the local level cannot necessarily be conceived of as purely local in a social context.

Problems that manifest at a local level, however conceived of, may not necessarily be able to be solved locally – even if local governance can play a part. And as even nature cannot be conceived of as "natural" in the meaning of being apart from society (Cronon 1996), understandings of nature, as well as what management and adaptations are developed, must be seen in social (including economic and political) and historical context. A true integration of socioecological perspectives – beyond socioecological theory – may require a better understanding of not only ecological but also social scale and processes.

The chapter has also illustrated that the local level is not necessarily less subject to what may have been seen as "large-scale" dynamics than higher levels are. The local level is impacted by multi-scalar dynamics, and interactions within these may equally need to be conceived of through historically developed interest structures, including economic structural interests and the legislative and policy environment in which they exist (and may have taken part in forming) (Keskitalo et al. 2016). "Stakeholders" or interests may be situated elsewhere as well, and the local level is impacted by organisations, legislation, and policy far beyond the local (and potentially difficult to change from a local perspective even if these were identified as "barriers" to change) (Pettersson et al. 2017).

The problem thus remains one of an inaccurate understanding of the complex social system: As noted earlier, barriers and limitations in dealing with a problem may not be removable per se without more extensive change, as they may be part of the organisational logics of, for instance, a sector, as issues are not single scale or necessarily possible to address at the local, or lower organisational, scale only – and problems may not even be possible to address as any single problem.

Our historical misconceptions of nature, society, small-scale, community, tradition, and related features can be seen here to play into the way in which specific areas have been highlighted more than others in relation to climate change and environmental study at large. Despite frequent criticism of these types of binary delineations of systems – nature-society, tradition-modernity, community-industry, and the like – they could be seen as living on despite being disproven: undead, just like the linear model of scientific knowledge (cf. Descola and Palsson 1996, eds). These issues may be seen as suggesting that it might be easier to formally confront an assumption like the need to deal with a separation between social and ecological thinking than to actually bring about change through practice, in existing institutions that acknowledge that social and natural science studies are established fields with accumulated knowledge that needs to be taken into account, rather than simply subsumed under more general conceptualisations that hold framings of their own (related to their origins and developers). As discussed throughout this book, many of the identified problems in fact result from seeing social systems, in a simplified way, as needing to reorganise to ecological organisation only, thus meaning that the issue of achieving a fit between ecological and social systems must also be seen entirely differently (Brulle and Dunlap 2015).

So as the chapter illustrates, there is no escaping the complexity of the social: In order to understand adaptation (or mitigation, or other) actions and possibilities, it is necessary to understand the constitution of interests, what they may be able to do, and how they can be incentivised otherwise. Particularly in relation to what has been discussed in this chapter, it is also necessary to understand how they are constituted through multiple levels – and in relation to conceptualisations of, for instance, tradition or community that may not match reality.

Key points

- The local level must be conceived of as constructed by historically developed practices and actions on all levels. The local level can be understood as part of an understanding of the legislative and policy framework, historical development, sectors, and interests that impact, for instance, the local level.
- Assuming the local to be a simpler case or treating it as such may only mean that you have omitted the more complex context that constitutes the local.

This could include legislation and policy, or even owners and interests that are not physically situated in or identifiable based on the locality.

- That the local level is, for instance, closer to nature or behaves in specific ways (such as being a community or being traditional) cannot be assumed but must rather be researched. Individuals can be traditional, close to nature or the like, but may be part of split, mixed, or several communities, where conflict or interests cannot be assumed to be absent.

Study questions

- How was the dichotomy of nature and society developed, and why does it persist to this day?
- Why is an assumption about community or the local as being closer to nature wrong?
- Why is the local necessarily part of a multilevel system and "not only local"?
- What would be the implications of seeing our understandings of what is natural and social as constructed (for instance, for debates on wilderness or rewilding)?

Notes

1 This idea that there would be "progress" from a more "primitive" to a more "advanced" stage in civilisational development has been described by, for instance, Kashima et al. (2011) as a "folk theory of social change" that has existed for a long time in various guises. It was not only a part of European colonialism, but also of the way powers long before described differences between what they tried to establish as "centre" versus "peripheries".

2 This is not to say that there is no literature on, for instance, social industrial relations; indeed, there is – but it is not the literature that has been placed in focus in adaptation work or in broader work focused on, for instance, socioecological systems (e.g. Olsson et al. 2004).

3 While forest owners' associations are less prominent in other cases (e.g. Keskitalo et al. 2017a), other historically developed structures, as well as the organization of land ownership, may have equally great importance. For instance, assumptions in the British estate-type ownership have come to also influence international conceptions of the countryside (even in the face of wide variety; Keskitalo et al. 2017b).

Additional readings

Nalau, J., Preston, B. L. and Maloney, M. C. (2015). Is adaptation a local responsibility? *Environmental Science & Policy* 48: 89–98.

Olsson, L., Jerneck, A., Thoren, H., Persson, J. and O'Byrne, D. (2015). Why resilience is unappealing to social science: Theoretical and empirical investigations of the scientific use of resilience. *Science Advances* 1(4): e1400217.

8

CONCLUSION

Implications of an institutional understanding

Introduction

This book has discussed the way in which social understandings relevant to environmental studies contrast with assumptions in the linear model of expertise or scientific knowledge. Focus has been placed on highlighting the implications of an institutional understanding using examples from multiple cases, particularly in forestry. The book has also specified the social focus with discussions of power and knowledge, contrasting Foucauldian understandings with a Habermasian understanding that may lie closer to assumptions in a linear model of scientific knowledge and discussing the implications on direct stakeholder participation as well as the roles of different scales. The book has also included a discussion on the historically based understandings of nature and society that may continue to impact how we conceive of environmental study and local levels even today.

The aim in this book has not been to simply provide one specific detailed definition of institutions and power, but rather to forward an understanding or framing of a way to conceive of the social. The book has been about forwarding an understanding, a conception of the social, that will hopefully prevent the reader from unquestioningly taking the linear model of scientific knowledge or a Habermasian ideal-type application to the real world as a given. Instead, the hope is that the reader will have internalised an understanding and interpretation of the complexity of the social as to realise that it, too, must be researched, in at least the same detail that climate change effects are, to at all begin to understand the implications, impacts, vulnerabilities, adaptation, mitigation considerations, and any other requirements entailed by a focus on climate or environmental change. What is more, it is also hoped that the reader will have understood that the types of considerations discussed and exemplified here apply not only to climate change but to all environmental issues – and furthermore, all types of issues in which social areas

DOI: 10.4324/9781003043867–8

are concerned. The social – conceived of as including political, economic, social, cultural, and other considerations – is not a black box (e.g. Wellstead et al. 2013) or quite simply "complex" or a "wicked" problem; it is an area where it is not only direct contact with selected stakeholders that is needed, but also an in-depth understanding of the institutions and power arrangements that create and constrain their circumstances and incentives, in order to support change.

To this end, this concluding chapter will focus on highlighting the implications of the very broad conceptualisations used in this book on how to understand institutions and power. The chapter will also look forward to one area that has been both present and absent in the book: that of integration among research areas.

Implications on social change from the book at large

The idea of "change" has often been seen as central to a focus on sustainability, and indeed as one way in which "sustainability science" is extended from a focus on science to a focus only on stakeholders (as discussed in Chapters 2 and 7). The general model of an environmental or sustainability science that operates along an understanding of natural science, added to mainly through participation, has been criticised throughout this book, as an example of how the linear model of scientific knowledge has been extended with a democratically oriented focus (cf. Dunlap and Brulle 2015, eds). However, the fact that possibilities and contexts for change have thereby been misconceived does not detract from the normative weight a broader science for sustainability may be given: Many who work in environmental areas are interested in broadly contributing to sustainability, and through this, different levels of change. What this book has hopefully done is contextualised the extent to which change is indeed not only a research problem for social studies but also fundamentally a political matter. Science does not itself steer political change but rather can only advise it.

On that basis, the focus here has been more on understanding the social system and comprehending it in a correct way, as broadly based in social science orientations. This itself can contribute to change, as it contradicts disproven assumptions concerning how change takes place, and may increase the possibility that change is problematised rather than taken as given (with some disappointment when it then does not occur).

In addition, the book has also highlighted means for social study to illustrate – in each case – the institutional logics that need to be understood in order to form a conception of where change could potentially be developed and, for instance, different policy instruments could be targeted (as discussed in Chapter 3). This latter part of selecting and targeting policy instruments is fundamentally a political issue: Science, including social science, can only advise. Hopefully, however, the discussion on policy instruments has also served to contextualise the extent to which they have to be adapted to specific aims based, for instance, on a social analysis rather than assumed as catch-all (given, for instance, the hopes that were earlier often attributed to new environmental governance mechanisms such as voluntary means).

As a whole, the book could thus be seen as problematising change. It has illustrated how social change in response to, for instance, climate change is circumscribed by social structures:

- It cannot be assumed that social change will take place simply based on better knowledge or that this knowledge will in an unfiltered way reach stakeholders, who will then learn and change. Instead, social change must be conceived of through an understanding of social structures; that is, one in which the necessities of change, understood through improved knowledge, need to be translated into incentives and motivations for change based on institutional context.
- Change can also never be assumed to be autonomous or automatic in response to knowledge but must be seen in the context of social institutions and an understanding of them.

Examples of this have been provided both in the discussion on institutions and in examples from forestry and national cases, highlighting that society is not necessarily easily shifted. For instance, the forestry case (discussed in Chapter 4, as well as further on) can be taken as one example of the type of understandings and studies that may be needed in other cases in order to understand the present structure and incentivisation. In this case, overarching institutions have been developed over centuries and include, for instance, the role of the state in its specific conceptualisation and its linkages to industry development and entrepreneurial development; the role of population and economic land use development in relation to multiple groups and interests; and how these processes result in an institutionalisation and to some extent a naturalisation of present systems. In the forestry case, for instance, change might need to relate to a shifting cost structure through governmental and potentially international agreements or limitations. For many other actors, given that all of them have at least an economic bottom line, many issues will be similar: How can an outcome be practically incentivised in relation to specific sustainability requirements?

While the fact of potentially strong institutionalisation in various areas does not at all preclude smaller changes and more limited institutions – but rather provides the context in which they are incentivised and developed – it does highlight the overarching role of understanding the specific historically developed institutions and how they can be incentivised, in order to "place into context the degree of change that may seem to occur at any given point" (Keskitalo 2019: 256). The question of how to achieve change is both that easy and that hard, as the agreements on specific sustainability requirements and their distribution across sectors may never be easy, and never interest-free. In addition, developing sustainability perspectives in detail will require both detailed institutional understanding and understandings that weigh in multiple sectors, areas, and interests, and the win-lose implications of any decisions.

So, for example, to understand why local change does not occur in any one given case, it may be necessary to also understand the state and different sectoral impacts on supporting or hindering this change, with an eye to the possibility that "supporting" or "hindering" may also be overly limited conceptions. Gaining any specific identified or desired change, for instance, may not so much be a question of simply having to "remove" "hindrances", but rather of understanding that they may be part of a larger system, to which they may be integral (as noted, for instance, in Chapters 2 and 7). Any one change can seldom be seen as separate but must instead be conceived of within the logic and incentive structure of the system – the potential paths and lock-ins, if you will (as noted in Chapter 3). While this book has focused on discussing incentivisation in terms of policy instruments, this does not mean that there are no other means for incentivising change: This focus is mainly used to illustrate the need to understand and work with people's motivations rather than going in blind with no relation to them. Various disciplines will no doubt describe certain types of incentivisation or motivation differently and will also (to differing extents) involve themselves with issues directly oriented towards change.

Thus, whether one argues that there is a need for transformation and more far-ranging rather than simply incremental change, or that transitions are the best ways to understand change, in the understanding forwarded in this book change is a question not so much of whether one prefers slow or fast, incremental or transformative change, but rather of how to develop the incentives for change that then determine how it can be enabled, based on an understanding of institutions (as discussed in Chapter 3).

The aim in this chapter is thus not to engage with additional bodies of literature or specific conceptualisations of change, which do exist in multiple variations. Any reader may consider what the implications of a focus on institutional dynamics might be for different bodies of literature and could, for instance, use the understanding here to assess such analyses on meta levels: Do they, for instance, relate to understanding of institutional dynamics that include power? Which of the types of issues discussed here are omitted from the frameworks or analyses? Which types of issues are emphasised or added?

Conceiving of change in this way, as an institutional issue based in analysis of real-life empirical systems, also results in implications on how we think about and study the future, and more so on how we think about development and research involving adaptation and mitigation. Future studies have often rested on scenario studies in which one foundational focus has been the target or goal: Towards what target do we aim, and what could then potentially make us miss this target or be adjusted so as to aim for it? While a great deal has been made of the fact that scenarios are not necessarily pathways that can necessarily be assessed as to how likely they are, less discussion has focused on the fact that scenarios place the crucial focus on assumptions regarding the future and future goals rather than on a deep-seated understanding of the present. As Beck (2011) argues, "the linear model of expertise places the scenario-based approach center-stage because it promises to deliver a 'sound' scientific foundation on which to base climate policy" (Beck 2011: 304). However, in relation to climate change, one might instead argue that by now we know the goals, for instance as they are set internationally. What has – interestingly enough, and not disregarding the traditions in the climate change field – been less of a focus is knowing the present: What are the institutional features that limit change towards this goal? How are specific sectors, countries, regions institutionally set up and what are the incentives that may be needed in order to shift business as usual?

As a whole, then:

> understanding the past (through conceptions of path dependencies, historical institutionalism, or related means) and the actors and motivations formed in relation to it may be one of the most sure ways of identifying actors that will determine decision-making paths in the future also, as well as what may drive their decisions under change. . . . To understand change, however, it is crucial to also understand that it is not one-dimensional, and that actors will need to respond to multiple stressors at the same time.
>
> *(Keskitalo 2019: 257)*

Thus, to understand the "future", the argument here has been that one must understand the present or even history: the ways in which embedded incentives have survived for long periods, and the ways in which different actors are incentivised and different paths of action encased in, for instance, legislative systems.

Must it be so hard? Addressing the role of science

Given the complexity of social systems that has been sketched in this book, it may be easy to understand why one would want to black-box social systems. Are there no certain inputs that always provide results? Can people not just agree and learn? Is there no magic bullet, no one generalisable solution?

But the answer is: No, there is no magic bullet. What is more, *the alternative to working with and attempting to understand the highly complex social systems is to not achieve change and not even start to comprehend why it does not occur.* For too long, climate change has been an area marked by specific scientific traditions – by power and institutions, in the very sense discussed in this book. It is a fact that social systems are not automatons whereby one input – not even the same instrument – can be made everywhere to gain the same effect, as previous chapters have illustrated. Each case requires analysis as to the types of factors highlighted here: the study of, rather than the exclusion of, specific interests and, for instance, resource-political contexts. However, this does not mean that social systems cannot be understood, or are too complex to be understood; it only means that there is a need for study and analysis rather than blanket statements and black-boxing. Taken together, much of this knowledge already exists as studies of specific organisations that will have a role in climate change adaptation and mitigation: We know how many international organisations and states work on numerous aspects, and there is various literature on different specific instruments and how they function under different conditions. The problem is that this type of knowledge has seldom been given a central role in relation to environmental and climate change studies in general.

In developing and supporting this understanding, science is central. The book at large has focused on illustrating why the linear model of scientific knowledge is wrong and on providing a social science-based conceptualisation of the types of phenomena made relevant (or obscured) by it. While the book has thereby focused on describing the social in relation to real-life empirical analysis relevant to, for instance, environmental issues, through this it has also addressed the need for scientific assessment and the broader debate on sustainability and climate change to correctly include these types of understandings. However, putting this type of understanding into practice would be no mean feat. The debate over sustainable development could be seen as a case in point: On sustainability, it has been noted that, despite significant research efforts organised around the concept of sustainable development, there are limitations in the extent to which these have generated major breakthroughs or resolved major intellectual and policy debates (Levy 2003; Gupta 2002; Keskitalo and Liljenfeldt 2012). Levy suggests that the problem is not so much the normative or teleological nature of the effort, or the woolly nature of the sustainability concept, as many research programmes are built on conceptually vague foundations (Levy 2003); instead, Levy notes, a part of the problem is that we sometimes think we are making progress, meeting data needs by filling separate thematic bins with relevant data and assuming that human and physical phenomena are possible to integrate in simplistic ways (Levy 2003).[1]

Dealing with issues of sustainability in research may require integration between what are – as we have seen in this volume – often highly different assumptions as to how the world works, or even what has to be researched or what is to be assumed. Understanding what has been discussed in previous chapters as implications of scale strongly relevant is in this case: As Levy notes, physical scientists may insist on pushing data requirements on physical phenomena to ever-increasing precision, but they are frequently content to use coarse social science data (Levy 2003). And while different assumptions may not come into contact with each other as long as different disciplines, subdisciplines, or assumptions about the world are kept apart, the issue of large problems that would require at least some coordination – or some way of gaining insights from all the areas of knowledge applicable to them – may thereby result in confusion rather than coherence or integration.

This begs another difficult question: What, then, are the possibilities for overcoming this divide and creating "true integration"? And should that be the aim?

In a book that contrasts different approaches relevant to a single disciplinary area, the conclusion is that the authors "find no easy way to combine a natural science approach with an interpretative one" (Hollis and Smith 1991: 6). They conclude: "Although it is appealing to believe that bits of the two stories can be added together, we maintain that there are always two stories to tell and that combinations do not solve the problem" (Hollis and Smith 1991: 7). What Hollis and Smith's (1991) conclusion relates itself to, though, is more the need to maintain both perspectives rather than assuming that one of them can suffice or tell the whole story by itself: There are always several stories to tell.

So perhaps the issue is not so much "integration" as it is as the need to allow all stories to be heard – to make visible all research relevant to the problem. The structure of science and research itself may be a large problem here. Problems in interdisciplinary communication relate not least to how problems and solutions are structured in the different disciplines, with a specificity that makes communication across disciplines difficult and with a risk of loss of face. It is a fact that as soon as you venture outside your discipline, you are no longer an expert; the questions you ask may seem stupid. Very few people are secure enough to both do this and be able to keep a discussion going to develop from this point. "Even scientists themselves may find it difficult to comprehend . . . results in specialities other than their own. Multidisciplinary teams then generally become a collection of specialists playing safe" (Funtowicz and Ravetz 1990: 60). And if "specialists playing safe" becomes the way in which framings are developed, based on who the dominant scientific players are – just like framings in society at large may be developed based on who the dominant players, actors, or organisations are – it may result in that not all stories are heard.

These things are regularly discussed in the theory of science. Some theorists of science even assert that what may be called a disciplinary blindness is a prerequisite for scientific organisation. Biagioli describes how, if scientific knowledge is the collective product of a group of interacting scientists, such a group needs to remain cohesive and committed to the articulation of its paradigm (Biagioli 1996). This "disciplinary blindness" produces not only social cohesion in scientific groups and the relative difficulty of communication across contexts; it also directs interests to some areas more than others – most of all, to disciplinary foci. These disciplinary foci will then resist challenge to their way of thinking. Management is often undertaken by avoidance, or by concentrating on known areas with developed interest. So, not only can knowledge be resisted because there are gaps in the data, dubious assumptions in the theoretical models, or countervailing evidence. It can also be resisted because the procedures and capabilities for its articulation and development are too sensitive, unprofitable, and so forth (Rouse 1996). As Rouse notes, this type of epistemic conflict or conflict around knowledge is always shaped by the goods, practices, and projects whose allocation and pursuit are at issue, and by the institutions and social networks that are organised around these pursuits (Rouse 1996) – i.e., by the institutional forces that have been the focus in this book. Research development is thus social and political: It is constituted by groups with their own preferences (Kuhn 1970; cf. Worster 1996), who do not necessarily incorrectly describe the things they do but do not describe all things. In emerging areas, such as that of global environmental change as a field, groups develop by drawing upon established disciplines and applying their tools to the new area. This does not mean that their tools are wrong or that the results that come out of their application are wrong, but it may mean that areas of research that do not select the same focus or apply the same tools are left out.

With this background, accomplishing integration around the same research area or even interaction becomes a tricky thing indeed. Disciplines exist because they limit and focus: This is what makes them disciplines. Trying to act upon complex social problems, or upon the issues of sustainable development, magnifies these problems extensively. It is not only society that is sectorial; science is as well. A general pattern of modern societies is functional differentiation in the departmentalisation of policy-making, the disciplinary split of scientific knowledge, and the branching of ever-more specialised professions. Differentiation, as Voss notes, is essentially a process of slicing up the world (Voss 2003). As Voss notes, the paradox here is that the fundamental motivation to develop knowledge about the world may turn against its own ends when it leads to ignorance about parts of reality (Voss 2003; cf. Konrad et al. 2004).

A case in point can be seen in the fact that the linear model of scientific knowledge, or different assumptions regarding how "communities" would necessarily work, remains in effect. This can be comprehended, for instance, through social theory. Examples include philosopher of science Kuhn's theory of scientific paradigms (that paradigms only die with the persons who carry them, Kuhn 1970) and Foucault's conception of discourses, as understandings that are so well linked to the groups that carry them that they are unable to think outside them (e.g. Foucault 1974). Based in this type of understanding, the main way to influence discourses is by having been schooled in different perspectives and being able to, and situated to, articulate them in relation to, for instance, dominating discourses. However, this does not mean that established discourses may necessarily change, because – as we have highlighted – better knowledge does not necessarily entail change, and the situatedness of actors can vary. As other theories express it, it would take a great deal to get your challenge on the agenda: involving not only you and associated interest groups being positioned for this, but also perhaps the confluence of several factors that make the problem you are targeting relevant to work on in relation to, for instance, events or other pressures (cf. Kingdon 1995).

Conclusions

If this book has been able to show one thing, it should hopefully be that social change is not simple. But what it should hopefully also have been able to show in this connection is that black-boxing social change, or assuming it will happen automatically, does not make it easier.

The way forward, then, for instance for you as a student or scholar reading this, may be to challenge yourself. You do not need to understand other sciences or become an interdisciplinary expert yourself, with expertise in multiple areas – even this book should have shown how much work that is. But what you would need to do is to assume that each discipline does what it does for some kind of reason;

that it has methodology and, for instance, theory (implicit or explicit); and that it generally knows what it is doing – with regard to its specific areas as they are treated disciplinarily, that is.

So if you are, for instance, a physical scientist, you should not assume that you know how to do stakeholder participation or even that this is the only thing that needs doing – even if this is the only social demand from the funding body. Instead, go to social science groups, and ask what they can provide and do regarding the question you are after. If you work on water, biodiversity, or climate change, perhaps they can work on the regulations surrounding these issues or on multilevel cases relevant to your area of focus. Anthropologists, geographers, and political scientists will all bring different understandings to a local case: Get them to battle it out and design their own ways to gain the knowledge they find relevant to the problem. If you are one among the 99.88% of funded climate change mitigation research internationally that is not social science (Overland and Sovacool 2020), find social scientists and delegate the responsibility for any social part of your research to them.

You do not need to know everything, but you need to know that the social is real, it is a topic of study, and people learn how to do this in disciplines: Let them do what their discipline dictates concerning the research problem area that you may have in common. Assume that everyone is an expert in their own disciplinary area. In this way, even if the institutions and traditions surrounding not only society but also science are as strong as we have seen them to be in this book, at least the biases of different sciences may balance each other and we will at least get results from different perspectives. And, with a stronger inclusion of a social sciences focus and a social understanding of what is needed for change, we will get more options developed in different socially grounded ways onto the table, and potentially further to decision-makers. This will not only be doable but will also at least be a progression from the present.

Change, transition, or transformation will thus, rather than only being big words, depend on what can be done in order to research and offer possibilities (and if you are a decision-maker, related incentives) for change. And what can be done needs to be understood in an institutional context as well as in contexts such as the existing literature on incremental versus catalytic change and the processes influencing this, or much other of the literature that has been reported throughout this book.

In this way, a final way of breaking with the linear model is to emphasise not only future targeted studies with a focus on scenario or other speculative approaches (cf. Beck 2011), but also research focused on understanding the past and present decision paths into the future, through institutional research, for instance. Understanding what can be done in the present, based on an understanding of institutional context, means studying not only the present but also how present actors and structures can be understood in order to potentially affect change or transformation into the future.

Key points

- A linear model of scientific knowledge and Habermasian assumptions regarding the social world can be contrasted with an institutional and power-centred perspective that is argued to be founded in what the world is actually like, rather than in ideal types of knowledge or knowledge transmission.
- Developing an understanding of possible change requires developing an understanding of the current real world, through social research. Researching this for one's case – as seen in a perspective of, among other things, scale and power – will provide the basis for understanding how any change can be incentivised.
- Science is a part of society – not apart from society – and is governed by the same processes that concern all of society. This does not mean that scientific knowledge is any less certain, but only that scientific knowledge in one area cannot cover other areas. Thus, natural science is not the authority on social processes; rather, social process knowledge must be seen as a research field within and be drawn from social science.
- While an institutional perspective on science as well may highlight difficulties of integration, the implication for the individual may be simple: Just assume that other disciplines are the experts on what they do and let them do it, bringing their focus to a research problem you may have in common. This in itself will be progress, moving away from assuming that the world works in ways (such as the linear model of scientific knowledge) that have not been empirically assessed by specific expertise in the given area.

Study questions

- How can we understand what the influences are regarding what change can be developed in any one social system case?
- Why do you think this text has placed so much focus on different disciplines?

Note

1 On local level initiatives, such as Agenda 21, it has been noted that the competition between different types of processes in some cases explained the limited results (e.g. Baker and Eckerberg 2007; cf. Lafferty and Cohen 2001). Implementation was seldom possible to conceive of in a simplistic way, or to see as automated. The difficulties were largely of the types that have been the focus in this work: that to become institutionalised, projects needed to fit within the institutions where they were to be implemented, and gain champions and leadership within the organisation as well as personnel and financial resources – in cases in which staff are often already overworked (Scheirer 2005; Da Silva and Shear 2010) and the competition with other, demanding, and already institutionalised issue areas is ever-present (e.g. Kingdon 1995; Keskitalo and Andersson 2017).

Additional readings

Dunlap, R. E. and Brulle, R. J. (eds.) (2015) *Climate Change and Society: Sociological Perspectives*. Oxford University Press, Oxford.

REFERENCES

Chapter 1

Beck, S. (2011) Moving beyond the linear model of expertise? IPCC and the test of adaptation. *Regional Environmental Change* 11(2): 297–306.

Beck, S. and Mahony, M. (2018) The IPCC and the new map of science and politics. *Wiley Interdisciplinary Reviews: Climate Change* 9(6): e547.

Böcher, M. and Krott, M. (2014) The RIU model as an analytical framework for scientific knowledge transfer: The case of the "decision support system forest and climate change". *Biodiversity and Conservation* 23(14): 3641–3656.

Briggle, A. (2008) Questioning expertise. *Social Studies of Science* 38(3): 461–470.

Brulle, R. J. and Dunlap, R. E. (2015) Sociology and climate change. In: Dunlap, R. E. and Brulle, R. J. (eds.) *Climate Change and Society: Sociological Perspectives*. Oxford University Press, Oxford. Pp. 1–21.

Collins, K. and Ison, R. (2009) Living with environmental change: Adaptation as social learning. *Environmental Policy and Governance* 19(6): 351–357.

De Koning, J., Turnhout, E., Winkel, G., Blondet, M., Borras, L., Ferranti, F., . . . Jump, A. (2014) Managing climate change in conservation practice: An exploration of the science – management interface in beech forest management. *Biodiversity and Conservation* 23(14): 3657–3671.

De la Vega-Leinert, A. C. and Schroter, D. (2009) Evaluation of a stakeholder dialogue on European vulnerability to global change. In: Patt, A., Schroter, D., Klein, R. and de la Vega-Leinert, A. (eds.) *Assessing Vulnerability to Global Environmental Change: Making Research Useful for Adaptation Decision Making and Policy*. Earthscan, London. Pp. 195–214.

De la Vega-Leinert, A. C., Schröter, D., Leemans, R., Fritsch, U. and Pluimers, J. (2008) A stakeholder dialogue on European vulnerability. *Regional Environmental Change* 8(3): 109–124.

Durant, D. (2015) The undead linear model of expertise. In: Heazle, M. and Kane, J. (eds.) *Policy Legitimacy, Science and Political Authority: Knowledge and Action in Liberal Democracies*. Routledge, London. Pp. 17–38.

Grundmann, R. (2009) The role of expertise in governance processes. *Forest Policy and Economics* 11(5–6): 398–403.

Hajer, M. A. (1995) *The Politics of Environmental Discourse: Ecological Modernization and the Policy Process*. Oxford University Press, Oxford.

Henderson, D. K. (1993) *Interpretation and Explanation in the Human Sciences*. SUNY Press, Albany.

Hollis, M. and Smith, S. (1991) *Explaining and Understanding International Relations*. Oxford University Press, Oxford.

IPCC (2014) *Climate Change 2014: Impacts, Adaptation, and Vulnerability. Part A: Global and Sectoral Aspects. Contribution of Working Group II to the Fifth Assessment Report of the Intergovernmental Panel on Climate Change*. Field, C. B., Barros, V. R., Dokken, D. J., Mach, K. J., Mastrandrea, M. D., Bilir, T. E., Chatterjee, M., Ebi, K. L., Estrada, Y. O., Genova, R. C., Girma, B., Kissel, E. S., Levy, A. N., MacCracken, S., Mastrandrea, P. R. and White, L. L. (eds.). Cambridge University Press, Cambridge, UK and New York. 1132 Pp.

IPCC AR4 WG2 (2007) In: Parry, M. L., Canziani, O. F., Palutikof, J. P., van der Linden, P. J. and Hanson, C. E. (eds.) *Climate Change 2007: Impacts, Adaptation and Vulnerability, Contribution of Working Group II to the Fourth Assessment Report of the Intergovernmental Panel on Climate Change*. Cambridge University Press, ISBN 978-0-521-88010-7 (pb: 978-0-521-70597-4).

Keskitalo, E. C. H. and Preston, B. L. (2019) Climate change adaptation policy research and its role in understanding climate change. In: Keskitalo, E. C. H. and Preston, B. L. (eds.) *Research Handbook on Climate Change Adaptation Policy*. Edward Elgar, Cheltenham. Pp. 475–491.

King, A. (2004) *The Structure of Social Theory*. London: Taylor & Francis.

Klein, R. J. T., Midgley, G. F., Preston, B. L., Alam, M., Berkhout, F. G. H., Dow, K. and Shaw, M. R. (2014) Adaptation opportunities, constraints, and limits. In: Field, C. B., Barros, V. R., Dokken, D. J., Mach, K. J., Mastrandrea, M. D., Bilir, T. E., Chatterjee, M., Ebi, K. L., Estrada, Y. O., Genova, R. C., Girma, B., Kissel, E. S., Levy, A. N., MacCracken, S., Mastrandrea, P. R. and White, L. L. (eds.) *Climate Change 2014: Impacts, Adaptation, and Vulnerability. Part A: Global and Sectoral Aspects. Contribution of Working Group II to the Fifth Assessment Report of the Intergovernmental Panel on Climate Change*. Cambridge University Press, Cambridge, UK and New York. Pp. 899–943.

Malone, E. L. and Engle, N. L. (2011) Evaluating regional vulnerability to climate change: Purposes and methods. *Wiley Interdisciplinary Reviews: Climate Change* 2(3): 462–474.

Manicas, P. T. (2006) *A Realist Philosophy of Social Science: Explanation and Understanding*. Cambridge University Press, Cambridge.

McAndrew, F. T. (1993) *Environmental Psychology*. Thomson Brooks/Cole Publishing Co.

Mimura, N., Pulwarty, R. S., Duc, D. M., Elshinnawy, I., Redsteer, M. H., Huang, H. Q., Nkem, J. N. and Sanchez Rodriguez, R. A. (2014) Adaptation planning and implementation. In: Field, C. B., Barros, V. R., Dokken, D. J., Mach, K. J., Mastrandrea, M. D., Bilir, T. E., Chatterjee, M., Ebi, K. L., Estrada, Y. O., Genova, R. C., Girma, B., Kissel, E. S., Levy, A. N., MacCracken, S., Mastrandrea, P. R. and White, L. L. (eds.) *Climate Change 2014: Impacts, Adaptation, and Vulnerability. Part A: Global and Sectoral Aspects. Contribution of Working Group II to the Fifth Assessment Report of the Intergovernmental Panel on Climate Change*. Cambridge University Press, Cambridge, UK and New York. Pp. 869–898.

Noble, I. R. (2019) The evolving interactions between adaptation research, international policy, and development practice. In: Keskitalo, E. C. H. and Preston, B. L. (eds.) *Research Handbook on Climate Change Adaptation Policy*. Edward Elgar, Cheltenham. Pp. 21–48.

Noble, I. R., Huq, S., Anokhin, Y. A., Carmin, J., Goudou, D., Lansigan, F. P., Osman-Elasha, B. and Villamizar, A. (2014) Adaptation needs and options. In: Field, C. B., Barros, V. R., Dokken, D. J., Mach, K. J., Mastrandrea, M. D., Bilir, T. E., Chatterjee, M., Ebi,

K. L., Estrada, Y. O., Genova, R. C., Girma, B., Kissel, E. S., Levy, A. N., MacCracken, S., Mastrandrea, P. R. and White, L. L. (eds.) *Climate Change 2014: Impacts, Adaptation, and Vulnerability. Part A: Global and Sectoral Aspects. Contribution of Working Group II to the Fifth Assessment Report of the Intergovernmental Panel on Climate Change*. Cambridge University Press, Cambridge, UK and New York. Pp. 833–868.

Olsson, L., Jerneck, A., Thoren, H., Persson, J. and O'Byrne, D. (2015) Why resilience is unappealing to social science: Theoretical and empirical investigations of the scientific use of resilience. *Science Advances* 1(4): e1400217.

Philips, A. (1995) *The Politics of Presence*. Claredon Press, Oxford.

Pielke, R. A. (2007) *The Honest Broker: Making Sense of Science in Policy and Politics*. Cambridge University Press, New York.

Proshansky, H. M., Ittelson, W. H. and Rivlin, L. G. (eds.) (1970) *Environmental Psychology: Man and His Physical Setting*. Holt, Rinehart and Winston, New York.

Reese, S. D. (2007) The framing project: A bridging model for media research revisited. *Journal of Communication* 57(1): 148–154.

Ruddin, L. P. (2006) You can generalize stupid! Social scientists, Bent Flyvbjerg, and case study methodology. *Qualitative Inquiry* 12(4): 797–812.

Schön, D. A. and Rein, M. (1994) *Frame Reflection. Toward the Resolution of Intractable Policy Controversies*. Basic Books, New York.

Snow, C. P. (1961) *The Two Cultures and the Scientific Revolution (Rede Lecture 1959)*. Cambridge University Press, New York.

Sovacool, B. K. (2014) What are we doing here? Analyzing fifteen years of energy scholarship and proposing a social science research agenda. *Energy Research & Social Science* 1: 1–29.

Stern, P. C. and Dietz, T. (2015) IPCC: Social scientists are ready. *Nature* 521: 161.

Tanner, T. (1980) Significant life experiences: A new research area in environmental education. *The Journal of Environmental Education* 11(4): 20–24.

Thornton, P. H., Ocasio, W. and Lounsbury, M. (2012) *The Institutional Logics Perspective. A New Approach to Culture, Structure and Process*. Oxford University Press, Oxford.

Victor, D. G. (2015) Embed the social sciences in climate policy. *Nature* 520: 27–29.

Von Storch, H., Bunde, A. and Stehr, N. (2011) The physical sciences and climate politics. In: Dryzek, J. S., Norgaard, R. B. and Schlosberg, D. (eds.) *The Oxford Handbook of Climate Change and Society*. Oxford University Press, Oxford. Pp. 113–128.

Weaver, D. H. (2007) Thoughts on agenda setting, framing, and priming. *Journal of Communication* 57(1): 142–147.

Wellstead, A. M., Howlett, M. and Rayner, J. (2013) The neglect of governance in forest sector vulnerability assessments: Structural-functionalism and "black box" problems in climate change adaptation planning. *Ecology and Society* 18(3): 23.

Young, O. R. (with contributions from Agrawal, A., King, L. A., Sand, P. H., Underdal, A. and Wasson, M., revised version prepared by H. Schroeder). (1999) *Institutional Dimensions of Global Environmental change. Science Plan*. IHDP, Bonn.

Zhou, Y. and Moy, P. (2007) Parsing framing processes: The interplay between online public opinion and media coverage. *Journal of Communication* 57(1): 79–98.

Chapter 2

Adger, W. N. (2006) Vulnerability. *Global Environmental Change* 16(3): 268–281.

Adger, W. N., Agrawala, S., Mirza, M. M. Q., Conde, C., O'Brien, K., Pulhin, J., Pulwarty, R., Smit, B. and Takahashi, K. (2007) Assessment of adaptation practices, options, constraints and capacity. In: Parry, M. L., Canziani, O. F., Palutikof, J. P., van der Linden, P. J. and Hanson, C. E. (eds.) *Climate Change 2007: Impacts, Adaptation and Vulnerability. Contribution of Working Group II to the Fourth Assessment Report of the Intergovernmental Panel on Climate Change*. Cambridge University Press, Cambridge, UK. Pp. 717–743.

Adger, W. N., Brown, K., Nelson, D. R., Berkes, F., Eakin, H., Folke, C., Galvin, K., Gunderson, L., Goulden, M., O'Brien, K., Ruitenbeek, J. and Tompkins, E. L. (2011) Resilience implications of policy responses to climate change. *WIREs Clim Change* 2: 757–766.

Antunes, P., Santos, R. and Videira, N. (2006) Participatory decision making for sustainable development – the use of mediated modelling techniques. *Land Use Policy* 23: 44–52.

Arts, B., Buizer, M., Horlings, L., Ingram, V., van Oosten, C. and Opdam, P. (2017) Landscape approaches: A state-of-the-art review. *Annual Review of Environment and Resources* 42: 439–463.

Bay-Larsen, I. and Hovelsrud, G. K. (2017) Activating adaptive capacities: Fishing communities in Northern Norway. In: Fondahl, G. and Wilson, G. N. (eds.) *Northern Sustainabilities: Understanding and Addressing Change in the Circumpolar World*. Springer, Dordrecht. Pp. 123–134.

Beck, S. (2011) Moving beyond the linear model of expertise? IPCC and the test of adaptation. *Regional Environmental Change* 11(2): 297–306.

Beck, S. and Mahony, M. (2018) The IPCC and the new map of science and politics. *Wiley Interdisciplinary Reviews: Climate Change* 9(6): e547.

Biesbroek, G. R., Termeer, C. J., Klostermann, J. E. and Kabat, P. (2014) Rethinking barriers to adaptation: Mechanism-based explanation of impasses in the governance of an innovative adaptation measure. *Global Environmental Change* 26: 108–118.

Briggle, A. (2008) Questioning expertise. *Social Studies of Science* 38(3): 461–470.

De la Vega-Leinert, A. C. and Schroter, D. (2009) Evaluation of a stakeholder dialogue on European vulnerability to global change. In: Patt, A., Schroter, D., Klein, R. and de la Vega-Leinert, A. (eds.) *Assessing Vulnerability to Global Environmental Change: Making Research Useful for Adaptation Decision Making and Policy*. Earthscan, London, UK. Pp. 195–214.

De la Vega-Leinert, A. C., Schröter, D., Leemans, R., Fritsch, U. and Pluimers, J. (2008) A stakeholder dialogue on European vulnerability. *Regional Environmental Change* 8(3): 109–124.

Durant, D. (2015) The undead linear model of expertise. In: Heazle, M. and Kane, J. (eds.) *Policy Legitimacy, Science and Political Authority: Knowledge and Action in Liberal Democracies*. Routledge, London. Pp. 17–38.

Ford, J. D., Stephenson, E., Cunsolo Willox, A., Edge, V., Farahbakhsh, K., Furgal, C., . . . Austin, S. (2016) Community-based adaptation research in the Canadian Arctic. *Wiley Interdisciplinary Reviews: Climate Change* 7(2): 175–191.

Hackmann, H., Moser, S. C. and Clair, A. L. S. (2014) The social heart of global environmental change. *Nature Climate Change* 4(8): 653.

Haider, W. and Morford, S. (2004) Relevance of social science to the management of natural resources in British Columbia. *BC Journal of Ecosystems and Management* 4(1): 1–11.

Hantula, J., Müller, M. M. and Uusivuori, J. (2014) Short communication international plant trade associated risks: Laissezfaire or novel solutions. *Environmental Science and Policy* 37: 158–160.

Holm, P., Goodsite, M. E., Cloetingh, S., Agnoletti, M., Moldan, B., Lang, D. J., . . . Scholz, R. W. (2013) Collaboration between the natural, social and human sciences in global change research. *Environmental Science & Policy* 28: 25–35.

Hovelsrud, G. K. and Smit, B. (eds.) (2010) *Community Adaptation and Vulnerability in Arctic Regions*. Springer, Dordrecht.

IPCC (2014) *Climate Change 2014: Impacts, Adaptation, and Vulnerability. Part A: Global and Sectoral Aspects. Contribution of Working Group II to the Fifth Assessment Report of the Intergovernmental Panel on Climate Change*. Field, C. B., Barros, V. R., Dokken, D. J., Mach, K. J., Mastrandrea, M. D., Bilir, T. E., Chatterjee, M., Ebi, K. L., Estrada, Y. O., Genova, R. C., Girma, B., Kissel, E. S., Levy, A. N., MacCracken, S., Mastrandrea, P. R. and White, L. L. (eds.). Cambridge University Press, Cambridge, UK and New York. 1132 Pp.

IPCC AR4 WG2 (2007) In: Parry, M. L., Canziani, O. F., Palutikof, J. P., van der Linden, P. J. and Hanson, C. E. (eds.) *Climate Change 2007: Impacts, Adaptation and Vulnerability, Contribution of Working Group II to the Fourth Assessment Report of the Intergovernmental Panel on Climate Change*. Cambridge University Press, ISBN 978-0-521-88010-7 (pb: 978-0-521-70597-4).

Kates, R. W. (1985) Preface. In: Kates, R. W., Ausubel, J. H. and Berberian, M. (eds.) *Climate Impact Assessment*. Wiley, New York. Pp xii–xix.

Keskitalo, E. C. H. (2004) A framework for multi-level stakeholder studies in response to global change. *Local Environment* (Special Issue on Multi-Level Governance) 9(5): 425–435.

Keskitalo, E. C. H. (2008) *Climate Change and Globalization in the Arctic: An Integrated Approach to Vulnerability Assessment*. Earthscan Publications, London. 257p.

Keskitalo, E. C. H. (2010) Conclusion: The development of adaptive capacity and adaptation measures in European countries. In: Keskitalo, E. C. H. (ed.) *Developing Adaptation Policy and Practice in Europe: Multi-level Governance of Climate Change*. Springer, Dordrecht. Pp. 339–366.

Keskitalo, E. C. H. (2011) How can forest management systems adapt to climate change? Possibilities in different forestry systems. *Forests* 2(1): 415–430.

Keskitalo, E. C. H., Dannevig, H., Hovelsrud, G. K., West, J. J. and Gerger Swartling, Å. (2011) Adaptive capacity determinants in developed states: Examples from the Nordic countries and Russia. *Regional Environmental Change* 11(3): 579–592.

Keskitalo, E. C. H., Horstkotte, T., Kivinen, S., Forbes, B. and Käyhkö, J. (2016) Generality of mis-fit? The real-life difficulty of matching scales in an interconnected world. *Ambio* 45(6): 742–752.

Keskitalo, E. C. H. and Preston, B. L. (2019a) Climate change adaptation policy research and its role in understanding climate change. In: Keskitalo, E. C. H. and Preston, B. L. (eds.) *Research Handbook on Climate Change Adaptation Policy*. Edward Elgar, Cheltenham. Pp. 475–491.

Keskitalo, E. C. H. and Preston, B. L. (2019b) Understanding adaptation in the context of social theory. In: Keskitalo, E. C. H. and Preston, B. L. (eds.) *Research Handbook on Climate Change Adaptation Policy*. Edward Elgar, Cheltenham. Pp. 2–20.

Keskitalo, E. C. H., Westerhoff, L. and Juhola, S. (2012) Agenda-setting on the Environment: The development of climate change adaptation as an issue in European States. *Environmental Policy and Governance* 22(6): 381–394.

Kingdon, J. W. (1995) *Agendas, Alternatives and Public Policies*. 2nd ed. HarperCollins, New York.

Klein, R. J. T., Midgley, G. F., Preston, B. L., Alam, M., Berkhout, F. G. H., Dow, K. and Shaw, M. R. (2014) Adaptation opportunities, constraints, and limits. In: Field, C. B., Barros, V. R., Dokken, D. J., Mach, K. J., Mastrandrea, M. D., Bilir, T. E., Chatterjee, M., Ebi, K. L., Estrada, Y. O., Genova, R. C., Girma, B., Kissel, E. S., Levy, A. N., MacCracken, S., Mastrandrea, P. R. and White, L. L. (eds.) *Climate Change 2014: Impacts,*

Adaptation, and Vulnerability. Part A: Global and Sectoral Aspects. Contribution of Working Group II to the Fifth Assessment Report of the Intergovernmental Panel on Climate Change. Cambridge University Press, Cambridge, UK and New York. Pp. 899–943.

Levin, K., Cashore, B., Bernstein, S. and Auld, G. (2012) Overcoming the tragedy of super wicked problems: Constraining our future selves to ameliorate global climate change. *Policy Sciences* 45(2): 123–152.

Malone, E. L. and Engle, N. L. (2011) Evaluating regional vulnerability to climate change: Purposes and methods. *Wiley Interdisciplinary Reviews: Climate Change* 2(3): 462–474.

Manicas, P. T. (2006) *A Realist Philosophy of Social Science: Explanation and Understanding.* Cambridge University Press, Cambridge.

Measham, T. G., Preston, B. L., Smith, T. F., Brooke, C., Gorddard, R., Withycombe, G. and Morrison, C. (2011) Adapting to climate change through local municipal planning: Barriers and challenges. *Mitigation and Adaptation Strategies for Global Change* 16: 889–909.

Mimura, N., Pulwarty, R. S., Duc, D. M., Elshinnawy, I., Redsteer, M. H., Huang, H. Q., Nkem, J. N. and Sanchez Rodriguez, R. A. (2014) Adaptation planning and implementation. In: Field, C. B., Barros, V. R., Dokken, D. J., Mach, K. J., Mastrandrea, M. D., Bilir, T. E., Chatterjee, M., Ebi, K. L., Estrada, Y. O., Genova, R. C., Girma, B., Kissel, E. S., Levy, A. N., MacCracken, S., Mastrandrea, P. R. and White, L. L. (eds.) *Climate Change 2014: Impacts, Adaptation, and Vulnerability. Part A: Global and Sectoral Aspects. Contribution of Working Group II to the Fifth Assessment Report of the Intergovernmental Panel on Climate Change.* Cambridge University Press, Cambridge, UK and New York. Pp. 869–898.

Montesclaros (2011) The Montesclaros declaration. Declaration from the participants of the International Union of Forest Research Organizations meeting held at the Montesclaros Monastery in Cantabria, Spain. May 23–27, 2011.

Moser, S. C. and Dilling, L. (2011) Communicating climate change: Closing the science-action gap. *The Oxford Handbook of Climate Change and Society*: 161–174.

Næss, L. O., Bang, G., Eriksen, S. and Vevatne, J. (2005) Institutional adaptation to climate change: Flood responses at the municipal level in Norway. *Global Environmental Change Part A* 15(2): 125–138.

Nalau, J., Preston, B. L. and Maloney, M. C. (2015) Is adaptation a local responsibility? *Environmental Science & Policy* 48: 89–98.

Noble, I. R. (2019) The evolving interactions between adaptation research, international policy, and development practice. In: Keskitalo, E. C. H. and Preston, B. L. (eds.) *Research Handbook on Climate Change Adaptation Policy*. Edward Elgar, Cheltenham. Pp. 21–48.

Noble, I. R., Huq, S., Anokhin, Y. A., Carmin, J., Goudou, D., Lansigan, F. P., Osman-Elasha, B. and Villamizar, A. (2014) Adaptation needs and options. In: Field, C. B., Barros, V. R., Dokken, D. J., Mach, K. J., Mastrandrea, M. D., Bilir, T. E., Chatterjee, M., Ebi, K. L., Estrada, Y. O., Genova, R. C., Girma, B., Kissel, E. S., Levy, A. N., MacCracken, S., Mastrandrea, P. R. and White, L. L. (eds.) *Climate Change 2014: Impacts, Adaptation, and Vulnerability. Part A: Global and Sectoral Aspects. Contribution of Working Group II to the Fifth Assessment Report of the Intergovernmental Panel on Climate Change*. Cambridge University Press, Cambridge, UK and New York. Pp. 833–868.

O'Brien, K. and Leichenko, R. (2000) Double exposure: Assessing the impacts of climate change within the context of economic globalization. *Global Environmental Change* 10: 221–232.

Oels, A. (2019) The promise and limits of participation in adaptation governance: Moving beyond participation towards disruption. In: Keskitalo, E. C. H. and Preston, B. L. (eds.) *Research Handbook on Climate Change Adaptation Policy*. Edward Elgar, Cheltenham. Pp. 138–156.

Overland, I. and Sovacool, B. K. (2020) The misallocation of climate research funding. *Energy Research & Social Science* 62: 101349.

Palsson, G., Szerszynski, B., Sörlin, S., Marks, J., Avril, B., Crumley, C., Hackmann, H., Holm, P., Ingram, J., Kirman, A., Pardo Buendía, M. and Weehuizen, R. (2013) Reconceptualizing the 'Anthropos' in the Anthropocene: Integrating the social sciences and humanities in global environmental change research. *Environmental Science & Policy* 28: 3–13.

Pedersen, D. B. (2016) Integrating social sciences and humanities in interdisciplinary research. *Palgrave Communications* 2(1): 1–7.

Pettersson, M., Strömberg, C. and Keskitalo, E. C. H. (2016) Possibility to implement invasive species control in Swedish forests. *Ambio* 45(2): 214–222.

Piattoni, S. (2009) Multi-level governance: A historical and conceptual analysis. *European Integration* 31(2): 163–180.

Pielke, R. A. (2007) *The Honest Broker: Making Sense of Science in Policy and Politics*. Cambridge University Press, New York.

Reed, J., Van Vianen, J., Deakin, E. L., Barlow, J. and Sunderland, T. (2016) Integrated landscape approaches to managing social and environmental issues in the tropics: Learning from the past to guide the future. *Global Change Biology* 22: 2540–2554.

Schön, D. A. and Rein, M. (1994) *Frame Reflection. Toward the Resolution of Intractable Policy Controversies*. Basic Books, New York.

Sovacool, B. K. (2014) What are we doing here? Analyzing fifteen years of energy scholarship and proposing a social science research agenda. *Energy Research & Social Science* 1: 1–29.

Stern, P. C. and Dietz, T. (2015) IPCC: Social scientists are ready. *Nature* 521: 161.

Tompkins, E. L. (2005) Planning for climate change in small islands: Insights from national hurricane preparedness in the Cayman Islands. *Global Environmental Change* 15(2): 139–149.

Victor, D. G. (2015) Embed the social sciences in climate policy. *Nature* 520: 27–29.

Watts, D. J. (2017) Should social science be more solution-oriented? *Nature Human Behaviour* 1(1): 15.

Wellstead, A. M., Howlett, M. and Rayner, J. (2013) The neglect of governance in forest sector vulnerability assessments: Structural-functionalism and "black box" problems in climate change adaptation planning. *Ecology and Society* 18(3): 23.

Wellstead, A. M., Rayner, J. and Howlett, M. (2014) Beyond the black box: Forest sector vulnerability assessments and adaptation to climate change in North America. *Environmental Science & Policy* 35: 109–116.

Young, O. R. (with contributions from Agrawal, A., King, L. A., Sand, P. H., Underdal, A. and Wasson, M., revised version prepared by H. Schroeder). (1999) *Institutional Dimensions of Global Environmental Change. Science Plan*. IHDP, Bonn.

Chapter 3

Alberini, A. and Segerson, K. (2002) Assessing voluntary programs to improve environmental quality. *Environmental and Resource Economics* 22(1–2): 157–184.

Baumgartner, F. R. and Jones, B. D. (1993) *Agendas and Instability in American Politics*. University of Chicago Press, Chicago.

Béland, D. (2005) Ideas, interests, and institutions: Historical institutionalism revisited. In: Lecours, A. (ed.) *New institutionalism: Theory and Analysis*. University of Toronto Press, Toronto. Pp. 29–50.

Bowen, F. (2014) *After Greenwashing: Symbolic Corporate Environmentalism and Society*. Cambridge University Press, Cambridge.

Burritt, R. L. and Welch, S. (1997) Australian commonwealth entities: An analysis of their environmental disclosures. *Abacus* 33: 69–87.

Cashore, B., Auld, G. and Newsom, D. (2004) *Governing through Markets – Forest Certification and the Emergence of Nonstate Authority*. Yale University Press, New Haven.

Cohen, M. D., March, J. G. and Olsen, J. P. (1972) A garbage can model of organizational choice. *Administrative Science Quarterly* 17(1): 1–25.

David, P. A. (1986) Understanding the Economics of QWERTY: The necessity of history. In: Parker, W. (ed.) *Economic History and the Modern Economist*. Basil Blackwell, Oxford.

Edelman, L. B. and Talesh, S. A. (2011) To comply or not to comply – That isn't the question: How organizations construct the meaning of compliance. In: Parker, C. and Lehmann Nielsen, V. (eds.) *Explaining Compliance: Business Responses to Regulation*. Edward Elgar, Cheltenham. Pp. 103–122.

Finnemore, M. and Toope, S. J. (2001) Alternatives to "legalization": Richer views of law and politics. *International Organization* 55(3): 743–758.

Flyvbjerg, B. (2006) Five misunderstandings about case-study research. *Qualitative Inquiry* 12(2): 219–245.

Foucault, M. (1974) *The Archaeology of Knowledge*. Tavistock, London.

Geels, F. W. (2011) The multi-level perspective on sustainability transitions: Responses to seven criticisms. *Environmental Innovation and Societal Transitions* 1: 24–40.

Giddens, A. (1984) *The Constitution of Society: Outline of the Theory of Structuration*. University of California Press, Berkeley.

Gillard, R., Gouldson, A., Paavola, J. and Van Alstine, J. (2016) Transformational responses to climate change: Beyond a systems perspective of social change in mitigation and adaptation. *Wiley Interdisciplinary Reviews: Climate Change* 7(2): 251–265.

Gingrich, J. (2015) Varying costs to change? Institutional change in the public sector. *Governance* 28(1): 41–60.

Gorges, M. J. (2001) New institutionalist explanations for institutional change: A note of caution. *Politics* 21(2): 137–145.

Greener, I. (2005) The potential of path dependence in political studies. *Politics* 25(1): 62–67.

Gunningham, N. (2009) Environment law, regulation and governance: Shifting architectures. *Journal of Environmental Law* 21: 179–212.

Harty, S. (2005) Theorizing institutional change. In: Lecours, A. (ed.) *New Institutionalism: Theory and Analysis*. University of Toronto Press, Toronto. Pp. 51–79.

Hollis, M. and Smith, S. (1991) *Explaining and Understanding International Relations*. Oxford University Press, Oxford.

Hooghe, L., Lenz, T. and Marks, G. (2019) *A Theory of International Organization*. Oxford University Press, Oxford.

Hooghe, L. and Marks, G. (2001) *Multi-level Governance and European Integration*. Rowman & Littlefield, Boulder, CO.

Hotho, J. and Saka-Helmhout, A. (2017) In and between societies: Reconnecting comparative institutionalism and organization theory. *Organization Studies* 38(5): 647–666.

Howlett, M. (2014) Why are policy innovations rare and so often negative? Blame avoidance and problem denial in climate change policy-making. *Global Environmental Change* 29: 395–403.

IPCC (2014) *Climate Change 2014: Impacts, Adaptation, and Vulnerability. Part A: Global and Sectoral Aspects. Contribution of Working Group II to the Fifth Assessment Report of the Intergovernmental Panel on Climate Change*. Field, C. B., Barros, V. R., Dokken, D. J., Mach, K.

J., Mastrandrea, M. D., Bilir, T. E., Chatterjee, M., Ebi, K. L., Estrada, Y. O., Genova, R. C., Girma, B., Kissel, E. S., Levy, A. N., MacCracken, S., Mastrandrea, P. R. and White, L. L. (eds.). Cambridge University Press, Cambridge, UK and New York. 1132 Pp.

IPCC AR4 WG2 (2007) In Parry, M. L., Canziani, O. F., Palutikof, J. P., van der Linden, P. J. and Hanson, C. E. (eds.) *Climate Change 2007: Impacts, Adaptation and Vulnerability, Contribution of Working Group II to the Fourth Assessment Report of the Intergovernmental Panel on Climate Change*. Cambridge University Press, ISBN 978-0-521-88010-7 (pb: 978-0-521-70597-4).

Jordan, A. R. K., Wurzel, W. and Zito, A. R. (2013) Still the century of 'new' environmental policy instruments? Exploring patterns of innovation and continuity. *Environmental Politics* 22(1): 155–173.

Jørgensen, T. B. (1999) The public sector in an in-between time: Searching for new public values. *Public Administration* 77: 565–584.

Kates, R. W. (1985) Preface. In: Kates, R. W., Ausubel, J. H. and Berberian, M. (eds.) *Climate Impact Assessment*. Wiley, New York. Pp xii–xix.

Kemp, R. and Soete, L. (1990) Inside the green box: On the economics of technological change the environment. In: Freeman, C. and Soete, L. (eds.) *New Explorations in the Economics of Technological Change*. Printer, London.

Keskitalo, E. C. H. and Andersson, E. (2017) Why organization may be the primary limitation to implementing sustainability at local level. Examples from Swedish case studies. *Resources* 6(1): 13ff.

Keskitalo, E. C. H., Horstkotte, T., Kivinen, S., Forbes, B. and Käyhkö, J. (2016) Generality of mis-fit? The real-life difficulty of matching scales in an interconnected world. *Ambio* 45(6): 742–752.

Keskitalo, E. C. H. and Liljenfeldt, J. (2012) Working with sustainability. Experiences of sustainability processes in Swedish municipalities. *Natural Resources Forum* 36: 16–27.

Keskitalo, E. C. H. and Liljenfeldt, J. (2014) Implementation of forest certification in Sweden: An issue of organisation and communication. *Scandinavian Journal of Forest Research* 29(5): 473–484.

Keskitalo, E. C. H. and Pettersson, M. (2016) Can adaptation to climate change at all be mainstreamed in complex multi-level governance systems? A case study of forest-relevant policies at the EU and Swedish levels. In: Leal Filho, W., Adamson, K., Dunk, R. M., Azeiteiro, U. M., Illingworth, S. and Alves, F. (eds.) *Implementing Climate Change Adaptation in Cities and Communities. Integrating Strategies and Educational Approaches*. Springer, Dordrecht. Pp. 53–74.

Keskitalo, E. C. H. and Preston, B. L. (2019a) Understanding adaptation in the context of social theory. In: Keskitalo, E. C. H. and Preston, B. L. (eds.) *Research Handbook on Climate Change Adaptation Policy*. Edward Elgar, Cheltenham. Pp. 2–20.

Keskitalo, E. C. H. and Preston, B. L. (2019b) Climate change adaptation policy research and its role in understanding climate change. In: Keskitalo, E. C. H. and Preston, B. L. (eds.) *Research Handbook on Climate Change Adaptation Policy*. Edward Elgar, Cheltenham. Pp. 475–491.

Keskitalo, E. C. H., Pettersson, M. and Sörlin, S. (2019) Introduction. Understanding historical contingencies into the future: Cases from northern Europe. In: Keskitalo, E. C. H. (ed.) *The Politics of Arctic Resources: Change and Continuity in the 'Old North' of Northern Europe*. Routledge, London and New York. Pp. 1–17.

Kingdon, J. W. (1995) *Agendas, Alternatives and Public Policies*. 2nd ed. HarperCollins, New York.

Kingston, C. and Caballero, G. (2009) Comparing theories of institutional change. *Journal of Institutional Economics* 5: 151–180.

Klein, R. J. T. (2011) Adaptation to climate change. In: Linkov, I. and Bridges, T. (eds.) *Climate*. NATO Science for Peace and Security Series C: Environmental Security. Springer, Dordrecht.

Klijn, E. H. and Koppenjan, J. F. M. (2000) Public management and policy networks. *Public Management* 2: 135–158.

Lafferty, W. M. and Cohen, F. (2001) Conclusions and perspectives. In: Lafferty, W. M. (ed.) *Sustainable Communities in Europe*. Earthscan, London.

Lakoff, G. and Johnson, M. (1980) *Metaphors We Live By*. University of Chicago Press, Chicago.

Le Manach, F., Jacquet, J. L., Bailey, M., Jouanneau, C. and Nouvian, C. (2020) Small is beautiful, but large is certified: A comparison between fisheries the Marine Stewardship Council (MSC) features in its promotional materials and MSC-certified fisheries. *PloS One* 15(5): e0231073.

Lenschow, A. (2014) Innovations through sector integration and new instruments. In: Lodge, M. and Wegrich, K. (eds.) *The Problem-solving Capacity of the Modern State: Governance Challenges and Administrative Capacities*. Oxford University Press, Oxford. Pp. 144–162.

Levy, J. D. (2015) State transformations in comparative perspective. In: Leibfried, S., Huber, E., Lange, M., Levy, J. D., Nullmeier, F. and Stephens, J. D. (eds.) *The Oxford Handbook of Transformations of the State*. Oxford University Press, Oxford. Pp. 169–190.

Liebowitz, S. J. and Margolis, S. E. (1995) Path dependence, lock-in, and history. *Journal of Law, Economics, & Organization*: 205–226.

Lindblom, C. E. (1959) The science of 'muddling through'. *Public Administration Review* 19: 78–88.

Lodge, M. (2014) Regulatory capacity. In: Lodge, M. and Wegrich, K. (eds.) *The Problem-solving Capacity of the Modern State: Governance Challenges and Administrative Capacities*. Oxford University Press, Oxford. Pp. 64–84.

Lubatkin, M. H., Lane, P. J., Collin, S. O. and Very, P. (2005) Origins of corporate governance in the USA, Sweden and France. *Organization Studies* 26(6): 867–888.

MacCormick, N. (2007) *Institutions of Law: An Essay in Legal Theory*. Oxford University Press, Oxford.

Mahoney, J. and Thelen, K. (2010) A theory of gradual institutional change. In: Mahony, J. and Thelen, K. (eds.) *Explaining Institutional Change. Ambiguity, Agency, and Power*. Cambridge University Press, New York.

Manicas, P. T. (2006) *A Realist Philosophy of Social Science: Explanation and Understanding*. Cambridge University Press, Cambridge.

Marara, M., Okello, N., Kuhanwa, Z., Douven, W., Beevers, L. and Leentvaar, J. (2011) The importance of context in delivering effective EIA: Case studies from East Africa. *Environmental Impact Assessment Review* 31(3): 286–296.

March, J. G. and Olsen, J. P. (1996) Institutional perspectives on political institutions. *Governance* 9(3): 247–264.

Marks, G. and Hooghe, L. (2004) Contrasting visions of multi-level governance. In: Bache, I. and Flinders, M. (eds.) *Multi-level Governance*. Oxford University Press, Oxford.

Marsden, G., Ferreira, A., Bache, I., Flinders, M. and Bartle, I. (2014) Muddling through with climate change targets: A multi-level governance perspective on the transport sector. *Climate Policy* 14(5): 617–636.

McDonald, J. (2011) The role of law in adapting to climate change. *WIREs Climate Change* 2: 283–295.

Meadowcroft, J. (2009) What about the politics? Sustainable development, transition management, and long term energy transition. *Policy Science* 42: 323–340.

Meadowcroft, J. (2011) Engaging with the politics of sustainable transitions. *Environmental Innovation and Societal Transitions* 1: 70–75.

Moran, E. (2010) *Environmental Social Science: Human-Environment Interactions and Sustainability*. New York: John Wiley & Sons.

North, D. C. (1990) *Institutions, Institutional Change and Economic Performance*. Cambridge University Press, Cambridge.

North, D. C. (1994) Institutional change: A framework of analysis. (Working paper republished). In: Braybrooke, D. (1998, ed.) *Social Rules: Origin; Character; Logic; Change*. Routledge, New York and London. Pp. 189–201.

Olsen, J. P. (2009) Change and continuity: An institutional approach to institutions of democratic government. *European Political Science Review: EPSR* 1(1): 3–32.

Olsson, L., Jerneck, A., Thoren, H., Persson, J. and O'Byrne, D. (2015) Why resilience is unappealing to social science: Theoretical and empirical investigations of the scientific use of resilience. *Science Advances* 1(4): e1400217.

Park, S. E., Marshall, N. A., Jakku, E., Dowd, A. M., Howden, S. M., Mendham, E. and Fleming, A. (2012) Informing adaptation responses to climate change through theories of transformation. *Global Environmental Change* 22(1): 115–126.

Piattoni, S. (2010) *The Theory of Multi-Level Governance: Conceptual, Empirical, and Normative Challenges*. Oxford University Press, Oxford.

Pierson, P. (2000) Increasing returns, path dependence, and the study of politics. *The American Political Science Review* 94(2): 251–267.

Potter, C. and Tilzey, M. (2005) Agricultural policy discourses in the European post-Fordist transition: Neoliberalism, neomercantilism and multifunctionality. *Progress in Human Geography* 29(5): 581–600.

Preston, B. L., Westaway, R. M. and Yuen, E. J. (2011) Climate adaptation planning in practice: An evaluation of adaptation plans from three developed nations. *Mitigation and Adaptation Strategies for Global Change* 16: 407–438.

Rosen-Zvi, I. (2011) You are too soft! What can corporate social responsibility do for climate change? *Minnesota Journal of Law, Science & Technology* 12: 527.

Ross, S. (2011) Spaces of play: Inventing the modern leisure space in British fiction and culture, 1860–1960. [Diss. English] Pennsylvania State University, Pennsylvania.

Ruddin, L. P. (2006) You can generalize stupid! Social scientists, Bent Flyvbjerg, and case study methodology. *Qualitative inquiry* 12(4): 797–812.

Runhaar, H., Driessen, P. and Uittenbroek, C. (2014) Towards a systematic framework for the analysis of environmental policy integration. *Environmental Policy and Governance* 24(4): 233–246.

Runhaar, H., Wilk, B., Persson, Å., Uittenbroek, C. and Wamsler, C. (2018) Mainstreaming climate adaptation: Taking stock about "what works" from empirical research worldwide. *Regional Environmental Change* 18(4): 1201–1210.

Sánchez, M. A. (2015) Integrating sustainability issues into project management. *Journal of Cleaner Production* 96: 319–330.

Sapotichne, J., Johnson, M. and Park, Y. S. (2013) Stability and change in US city policymaking: Evidence and a path forward. *Urban Research & Practice* 6(3): 255–275.

Smith, J. B., Vogel, J. M. and Cromwell III, J. E. (2009) An architecture for government action on adaptation to climate change. An editorial comment. *Climatic Change* 95: 53–61.

Sørensen, E. and Torfin, J. (2014) Collaborative innovation and governance capacity. In: Lodge, M. and Wegrich, K. (eds.) *The Problem-solving Capacity of the Modern State: Governance Challenges and Administrative Capacities*. Oxford University Press, Oxford. Pp. 239–258.

Sovacool, B. K., Ryan, S. E., Stern, P. C., Janda, K., Rochlin, G., Spreng, D., . . . Lutzenhiser, L. (2015) Integrating social science in energy research. *Energy Research & Social Science* 6: 95–99.

Streeck, W. and Thelen, K. (eds.) (2005) *Beyond Continuity: Institutional Change in Advanced Political Economies*. Oxford University Press, Oxford.

Stringer, L. C., Dougill, A. J., Fraser, E., Hubacek, K., Prell, C. and Reed, M. S. (2006) Unpacking "participation" in the adaptive management of social – ecological systems: A critical review. *Ecology and Society* 11(2): 39.

Talbot, C. (2000) Performing 'performance' – A comedy in five acts. *Public Money Management* 20: 63–68.

Thornton, P. H., Ocasio, W. and Lounsbury, M. (2012) *The Institutional Logics Perspective. A New Approach to Culture, Structure and Process*. Oxford University Press, Oxford.

Tsarouhas, D. and Ladi, S. (2013) Globalisation and/or Europeanisation? The case of flexicurity. *New Political Economy* 18(4): 480–502.

Van Bommel, K. (2014) Towards a legitimate compromise? An exploration of integrated reporting in the Netherlands. *Accounting, Auditing & Accountability Journal* 27(7): 1157–1189.

Van den Bergh, J. C. (2013) Environmental and climate innovation: Limitations, policies and prices. *Technological Forecasting and Social Change* 80(1): 11–23.

Victor, D. G. (2015) Embed the social sciences in climate policy. *Nature* 520: 27–29.

Walz, A., Lardelli, C., Behrendt, H., Grêt-Regamey, A., Lundström, C., Kytzia, S. and Bebi, P. (2007) Participatory scenario analysis for integrated regional modelling. *Landscape and Urban Planning* 81(1–2): 114–131.

Wellstead, A. M., Howlett, M. and Rayner, J. (2013) The neglect of governance in forest sector vulnerability assessments: Structural-functionalism and "black box" problems in climate change adaptation planning. *Ecology and Society* 18(3): 23.

Young, O. R. (with contributions from Agrawal, A., King, L. A., Sand, P. H., Underdal, A. and Wasson, M., revised version prepared by H. Schroeder). (1999) *Institutional Dimensions of Global Environmental Change. Science Plan*. IHDP, Bonn.

Zahariadis, N. (1997) Ambiguity, time, and multiple streams. In: Sabatier, P. (ed.) *Theories of the Policy Process*. Westview Press, Boulder, CO. Pp. 73–93.

Chapter 4

Andersson, E. and Keskitalo, E. C. H. (2018) Adaptation to climate change? Why business-as-usual remains the logical choice in Swedish forestry. *Global Environmental Change* 48(1): 76–85.

Andersson, E. and Keskitalo, E. C. H. (2019) Service logics and strategies of Swedish forestry in the structural shifts of forest ownership: Challenging the "old" and shaping the "new". *Scandinavian Journal of Forest Research* 34(6): 508–520.

Andersson, E., Keskitalo, E. C. H. and Bergstén, S. (2018) In the eye of the storm: Adaptation logics of forest owners in management and planning in Swedish areas. *Scandinavian Journal of Forest Research* 33(8): 800–808.

Andersson, E., Keskitalo, E. C. H. and Westin, K. (2020) Managing place and distance: Restructuring sales and work relations to meet urbanisation-related challenges in Swedish forestry. *Forest Policy and Economics* 118: 1–7.

Andersson, L., Bohman, A., van Well, L., Jonsson, A., Persson, G. and Farelius, J. (2015) *Underlag till kontrollstation 2015 för anpassning till ett förändrat klimat*. SMHI, Norrköping.

Appelstrand, M. (2007) *Miljömålet i skogsbruket-styrning och frivillighet*. Lund University, Lund.

Axelsson, A.-L. and Östlund, L. (2001) Retrospective gap analysis in a Swedish boreal forest landscape using historical data. *Forest Ecology and Management* 147(2–3): 109–122.

Axelsson, P., Sköld, P. and Röver, C. (2019) Ethnic identity and resource rights in Sweden. In: Keskitalo, E. C. H. (ed.) *The Politics of Arctic Resources: Change and Continuity in the 'Old North' of Northern Europe*. Routledge, London and New York. Pp. 119–139.

Back, A. and Marjavaara, R. (2017) Mapping an invisible population: The uneven geography of second-home tourism. *Tourism Geographies* 19(4): 595–611.

Bay-Larsen, I. and Hovelsrud, G. K. (2017) Activating adaptive capacities: Fishing communities in Northern Norway. In: Fondahl, G. and Wilson, G. N. (eds.) *Northern sustainabilities: Understanding and Addressing Change in the Circumpolar World*. New York: Springer. Pp. 123–134.

Beck, S. (2011) Moving beyond the linear model of expertise? IPCC and the test of adaptation. *Regional Environmental Change* 11(2): 297–306.

COM (2013) 216 final (2013) Communication from the commission to the European parliament, the council, the European economic and social committee and the committee of the regions. An EU Strategy on Adaptation to Climate Change. http://eur-lex.europa.eu/LexUriServ/LexUriServ.do?uri¼CELEX:DKEY¼725522:EN:NOT.

Commission on Climate and Vulnerability (2007) *Sweden Facing Climate Change: Threats and Opportunities: Final Report (SOU 2007:60)*. Swedish Government Official Report, Stockholm.

Ellison, D. (2010) Addressing adaptation in the EU policy framework. In: Keskitalo, E. C. H. (ed.) *Developing Adaptation Policy and Practice: Multi-level Governance of Climate Change*. Springer, Berlin.

Gordon, C. (1991) Governmental rationality: An introduction. In: Foucault, M., Burchell, G. and Gordon, C. (eds.) *The Foucault Effect: Studies in Governmentality: With Two Lectures by and an Interview with Michel Foucault*. University of Chicago Press, Chicago. Pp. 1–51.

Government Offices of Sweden (2009) *An Integrated Climate and Energy Policy* (Prop.2008/09:162–163). Swedish Government Bills, Stockholm.

Government Offices of Sweden (2017) *Klimatlag (SFS 2017:720)*. Swedish Government Bills, Stockholm.

Government Offices of Sweden (2018) *Nationell strategi för klimatanpassning* (Prop.2017/18:163). Government Offices of Sweden, Stockholm.

Haugen, K., Karlsson, S. and Westin, K. (2016) New forest owners: Change and continuity in the characteristics of Swedish non-industrial private forest owners (NIPF Owners) 1990–2010. *Small-scale Forestry* 15: 533–550.

Hysing, E. (2009) Governing without government? The private governance of forest certification in Sweden. *Public Administration* 87(2): 312–326.

Johansson, J. and Keskitalo, E. C. H. (2014) Coordinating and implementing multiple systems for forest management: Implications of the regulatory framework for sustainable forestry in Sweden. *Journal of Natural Resources Policy Research* 6(2–3): 117–133.

Juhola, S., Keskitalo, E. C. H. and Westerhoff, L. (2011) Understanding the framings of climate change adaptation across multiple scales of governance in Europe. *Environmental Politics* 20(4): 445–463.

Keskitalo, E. C. H. (2010a) Adapting to climate change in Sweden: National policy development and adaptation measures in Västra Götaland. In: Keskitalo, E. C. H. (ed.) *Developing Adaptation Policy and Practice in Europe: Multi-level Governance of Climate Change*. Springer, Dordrecht. Pp. 189–232.

Keskitalo, E. C. H. (2010b) Climate change adaptation in the United Kingdom: England and South-East England. In: Keskitalo, E. C. H. (ed.) *Developing Adaptation Policy and Practice in Europe: Multi-level Governance of Climate Change*. Springer, Dordrecht. Pp. 97–147.

Keskitalo, E. C. H. (ed.) (2010c) *Developing Adaptation Policy and Practice in Europe: Multi-level Governance of Climate Change*. Springer, Dordrecht. 379p.

Keskitalo, E. C. H. (2011) How can forest management systems adapt to climate change? Possibilities in different forestry systems. *Forests* 2(1): 415–430.

Keskitalo, E. C. H. (2017) Introduction. In: Keskitalo, E. C. H. (ed.) *Globalisation and Change in Forest Ownership and Forest Use: Natural Resource Management in Transition*. Palgrave Macmillan, Basingstoke. Pp. 1–16.

Keskitalo, E. C. H. (2008a) Konflikter mellan rennäring och skogsbruk i Sverige. In: Sandström, C., Hovik, S. and Falleth, E. I. (eds.) *Omstridd natur. Trender & utmaningar i nordisk naturförvaltning*. Borea, Umeå. Pp. 248–268.

Keskitalo, E. C. H. (2008b) *Climate Change and Globalization in the Arctic: An Integrated Approach to Vulnerability Assessment*. Earthscan Publications, London. 257p.

Keskitalo, E. C. H., Bergh, J., Felton, A., Björkman, C., Berlin, M., Axelsson, P., Ring, E., Ågren, A., Roberge, J.-M., Klapwijk, M. J. and Boberg, J. (2016) Adaptation to climate change in Swedish forestry. *Forests* 7: 28.

Keskitalo, E. C. H., Eklöf, J. and Nordlund, C. (2011) Climate change mitigation and adaptation in Swedish forests: Promoting forestry, capturing carbon and Fuelling transports. In: Järvelä, M. and Juhola, S. (eds.) *Energy and the Environment in the North – Competing Powers?* Springer, Dordrecht.

Keskitalo, E. C. H., Juhola, S. and Westerhoff, L. (2012b) Climate change adaptation as governmentality: Technologies of government in the development of adaptation policy in four countries. *Journal of Environmental Planning and Management* 55(4): 1–18.

Keskitalo, E. C. H., Karlsson, S., Lindgren, U., Pettersson, Ö., Lundmark, L., Slee, B., Villa, M. and Feliciano, D. (2017a) Rural – urban politics: Changing conceptions of the human-environment relationship. In: Keskitalo, E. C. H. (ed.) *Globalisation and Change in Forest Ownership and Forest Use: Natural Resource Management in Transition*. Palgrave Macmillan, Basingstoke. Pp. 183–224.

Keskitalo, E. C. H., Lidestav, G., Karppinen, H. and Zivojinovic, I. (2017b) Is there a new European forest owner? The institutional context. In: Keskitalo, E. C. H. (ed.) *Globalisation and Change in Forest Ownership and Forest Use: Natural Resource Management in Transition*. Palgrave Macmillan, Basingstoke. Pp. 17–56.

Keskitalo, E. C. H., Nocentini, S. and Bottalico, F. (2013) Adaptation to climate change in forest management: What role does national context and forest management tradition play? In: Manuel Esteban Lucas-Borja (ed.) *Forest Management of Mediterranean Forest Under the New Context of Climate Change*. Nova Science Publishers, New York. Pp. 149–161.

Keskitalo, E. C. H. and Pettersson, M. (2012) 'Implementing multi-level governance? The legal basis and implementation of the EU water framework directive for forestry in Sweden. *Environmental Policy and Governance* 22(2), March/April: 90–103. Re-printed as Chapter 35 in Bache, I. and Flinders, M. (2015) *Multi-Level Governance: Essential Readings*. Elgar Research Reviews in Social and Political Science. Edward Elgar Press, Cheltenham.

Keskitalo, E. C. H. and Pettersson, M. (2016) Can adaptation to climate change at all be mainstreamed in complex multi-level governance systems? A case study of forest-relevant policies at the EU and Swedish levels. In: Leal Filho, W., Adamson, K., Dunk, R. M., Azeiteiro, U. M., Illingworth, S. and Alves, F. (eds.) *Implementing Climate Change Adaptation in Cities and Communities. Integrating Strategies and Educational Approaches*. Springer, Dordrecht. Pp. 53–74.

Keskitalo, E. C. H., Westerhoff, L. and Juhola, S. (2012a) Agenda-setting on the environment: The development of climate change adaptation as an issue in European States. *Environmental Policy and Governance* 22(6): 381–394.

Kingdon, J. W. (1995) *Agendas, Alternatives and Public Policies*. 2nd ed. HarperCollins, New York.

Klapwijk, M. J., Boberg, J., Bergh, J., Bishop, K., Björkman, C., Ellison, D., Felton, A., Lidskog, R., Lundmark, T., Keskitalo, E. C. H., Sonesson, J., Nordin, A., Nordström, E.-M., Stenlid, J. and Mårald, E. (2018) Capturing complexity: Forests, decision-making and climate change mitigation action. *Global Environmental Change* 52: 238–247.

Klein, R. J. T. (2011) Adaptation to climate change. In: Linkov, I. and Bridges, T. (eds.) *Climate*. NATO Science for Peace and Security Series C: Environmental Security. Springer, Dordrecht.

Kunnas, J., Keskitalo, E. C. H., Pettersson, M. and Stjernström, O. (2019) The institutionalisation of forestry as a primary land use in Sweden. In: Keskitalo, E. C. H. (ed.) *The Politics of Arctic Resources: Change and Continuity in the 'Old North' of Northern Europe*. Routledge, London and New York. Pp. 62–77.

Kurz, W. A., Beukema, S. J. and Apps, M. J. (1997) Carbon budget implications of the transition from natural to managed disturbance regimes in forest landscapes. *Mitigation and Adaptation Strategies for Global Change* 2(4): 405–421.

Kurz, W. A., Stinson, G. and Rampley, G. (2008) Could increased boreal forest ecosystem productivity offset carbon losses from increased disturbances? *Philosophical Transactions of the Royal Society* 363(1501): 2259–2268.

Lazlo Ambjörnsson, E., Keskitalo, E. C. H. and Karlsson, S. (2016) Forest discourses and the role of planning-related perspectives: The case of Sweden. *Scandinavian Journal of Forest Research* 31(1): 111–118.

Lidestav, G., Thellbro, C., Sandström, P., Lind, T., Holm, E., Olsson, O., Westin, K., Karppinen, H. and Ficko, A. (2017) Interactions between forest owners and their forests. In: Keskitalo, E. C. H. (ed.) *Globalisation and Change in Forest Ownership and Forest Use: Natural Resource Management in Transition*. Palgrave Macmillan, Basingstoke. Pp. 97–137.

Lönnstedt, L. (2014) Swedish forest owners' associations: Establishment and development after the 1970s. *Small-scale Forestry* 13: 219–235.

Lundmark, T., Bergh, J., Hofer, P., Lundström, A., Nordin, A., Poudel, B. C., . . . Werner, F. (2014) Potential roles of Swedish forestry in the context of climate change mitigation. *Forests* 5(4): 557–578.

Marks, G. and Hooghe, L. (2004) Contrasting visions of multi-level governance. In: Bache, I. and Flinders, M. (eds.) *Multi-level Governance*. Oxford University Press, Oxford.

McDermott, C. L., Cashore, B. and Kanowski, P. (2010) *Global Environmental Forest Policies: An International Comparison*. Earthscan, London.

Öberg, P., Svensson, T., Christiansen, P. M., Nørgaard, A. S., Rommetvedt, H. and Thesen, G. (2011) Disrupted exchange and declining corporatism: Government authority and interest group capability in Scandinavia. *Government and Opposition* 46(3): 365–391.

Ogilvie, S. and Cerman, M. (eds.) (1996) *European Proto-Industrialization: An Introductory Handbook*. Cambridge University Press, Cambridge.

Parviainen, J. (2006) Forest management and cultural heritage. In: MCotPoFi Europe (ed.) *Forest and Our Cultural Heritage: Proceedings for the Seminar 13–15 June 2005, Sunne, Sweden*. Liaison Unit, Warsaw. Pp. 67–75.

Pettersson, M., Stjernström, O. and Keskitalo, E. C. H. (2017) The role of participation in the planning process: Examples from Sweden. *Local Environment* 22(8): 986–997.

Pierre, J. and Peters, B. G. (2005) *Governing Complex Societies – Trajectories and Scenarios*. Palgrave MacMillan, Basingstoke.

Rose, N. (1996) Governing 'advanced' liberal democracies. In: Barry, A., Osborne, T. and Rose, N. (eds.) *Foucault and Political Reason: Liberalism, Neo-liberalism, and Rationalities of Government*. University of Chicago Press, Chicago. Pp. 37–64.

Rose, N. and Miller, P. (1992) Political power beyond the state: Problematics of government. *British Journal of Sociology* 43(2): 173–205.

Rye, J. F. and Gunnerud Berg, N. (2011) The second home phenomenon and Norwegian rurality, *Norsk Geografisk Tidsskrift – Norwegian Journal of Geography* 65(3): 126–136.

SFIF [Swedish Forest Industries Federation] (2020) Facts & figures. Sweden's forest industry in brief. www.forestindustries.se/forest-industry/facts-and-figures/ (Accessed July 22, 2020).

Stjernström, O., Ahas, R., Bergstén, S., Eggers, J., Hain, H., Karlsson, S., Keskitalo, E. C. H., Lämås, T., Pettersson, Ö., Sandström, P. and Öhman, K. (2017) Multi-level planning and conflicting interests in the forest landscape. In: Keskitalo, E. C. H. (ed.) *Globalisation and Change in Forest Ownership and Forest Use: Natural Resource Management in Transition*. Palgrave Macmillan, Basingstoke. Pp. 225–260.

SWD/2013/131 final (2013) Commission staff working document. Summary of the impact assessment accompanying the document communication an EU strategy on adaptation to climate change. http://eur-lex.europa.eu/LexUriServ/LexUriServ.do?uri¼CELEX:52013SC0131:EN:NOT.

SWD (2013) 132 final (2013) Commission staff working document. Impact assessment – Part 2. Accompanying the document Communication from the Commission to the European Parliament, the Council, the European Economic and Social Committee and the Committee of the Regions. An EU Strategy on adaptation to climate change. http://ec.europa.eu/clima/ policies/adaptation/what/docs/swd_2013_132_2_en.pdf 788.

Törnqvist, T. (1995) *Skogsrikets arvingar: en sociologisk studie av skogsägarskapet inom privat, enskilt skogsbruk*. Swedish University of Agricultural Sciences, Uppsala.

Vepsäläinen, M. and Pitkänen, K. (2010) Second home countryside. Representations of the rural in Finnish popular discourses. *Journal of Rural Studies* 26: 194–204.

Westerhoff, L., Keskitalo, E. C. H. and Juhola, S. (2011) Capacities across scales: Local to national adaptation policy in four European countries. *Climate Policy* 11(4): 1071–1085.

Winkel, G. (2012) Foucault in the forests – A review of the use of 'Foucauldian' concepts in forest policy analysis. *Political Economy of Economic* 16: 81–92.

Chapter 5

Allen, A. (2010) The entanglement of power and validity: Foucault and critical theory. In: O'Leary, T. and Falzon, C. (eds.) *Foucault and Philosophy*. Wiley Blackwell, Chichester. Pp. 78–98.

Allen, B. (2010) Foucault's theory of knowledge. In: O'Leary, T. and Falzon, C. (eds.) *Foucault and Philosophy*. Wiley Blackwell, Chichester. Pp. 143–161.

Ashenden, S. and Owen, D. (eds.) (1999a) *Foucault Contra Habermas: Recasting the Dialogue between Genealogy and Critical Theory*. Sage Publications, London.

Ashenden, S. and Owen, D. (1999b) Introduction: Foucault, Habermas and the politics of critique. In: Ashenden, S. and Owen, D. (eds.) *Foucault Contra Habermas: Recasting the Dialogue between Genealogy and Critical Theory*. Sage Publications, London.

Barnett, M. and Duvall, R. (2005) Power in international politics. *International Organization* 59(1): 39–75.

Cashore, B., Auld, G. and Newsom, D. (2004) *Governing Through Markets – Forest Certification and the Emergence of Nonstate Authority*. Yale University Press, New Haven.

Dean, M. (1999) Normalising democracy: Foucault and Habermas on democracy. In: Ashenden, S. and Owen, D. (eds.) *Foucault Contra Habermas: Recasting the Dialogue between Genealogy and Critical Theory*. Sage Publications, London.

Dreyfus, H. L. and Rabinow, P. (2014) *Michel Foucault: Beyond Structuralism and Hermeneutics*. University of Chicago Press, Chicago.

Dryzek, J. S. (1990) *Discursive Democracy: Politics, Policy, and Political Science*. Cambridge University Press, Cambridge.

Eagly, A. H. and Kulesa, P. (1997) Attitudes, attitude structure, and resistance to change. Implications for persuasion on environmental issues. In: Bazerman, M. H. et al. (eds.) *Environment, Ethics and Behavior. The Psychology of Environmental Valuation and Degradation*. The New Lexington Press, San Francisco.

Eley, G. (1992) Nations, publics, and political cultures: Placing Habermas in the Nineteenth century. In: Calhoun, C. (ed.) *Habermas and the Public Sphere*. MIT Press, Cambridge, MA.

Foucault, M. (1973) *The Order of Things: An Archaeology of the Human Sciences*. Vintage, New York.

Foucault, M. (1974) *The Archaeology of Knowledge*. Tavistock, London.

Gentner, D. and Whitley, E. W. (1997) Mental models of population growth. A preliminary investigation. In: Bazerman, M. H. et al. (eds.) *Environment, Ethics and Behavior. The Psychology of Environmental Valuation and Degradation*. The New Lexington Press, San Francisco.

Gorges, M. J. (2001) New institutionalist explanations for institutional change: A note of caution. *Politics* 21(2): 137–145.

Habermas, J. (1984) *The Theory of Communicative Action, vol. 1: Reason and the Rationalization of Society*. Beacon Press, Boston.

Habermas, J. (1987) *The Theory of Communicative Action, vol. 2: Lifeworld and System: A Critique of Functionalist Reason*. Beacon Press, Boston.

Habermas, J. (1996) *Between Facts and Norms: Contributions to a Discourse Theory of Law and Democracy*. Polity Press, Cambridge.

Hadden, S. G. (1995) Regulatory negotiation as citizen participation: A critique. In: Renn, O., Webler, T. and Wiedermann, P. (eds.) *Fairness and Competence in Citizen Participation. Evaluating Models for Environmental Discourse*. Kluwer Academic Publishers, Dordrecht and London.

Holmes, T. and Scoones, I. (2000) Participatory Environmental Policy Processes: Experiences from North and South. IDS Working Paper 113. Institute of Development Studies, Brighton.

Ingram, D. (2005) Foucault and Habermas. In: Gutting, G. (ed.) *The Cambridge Companion to Foucault*. Cambridge University Press, Cambridge. Pp. 240–283.

Keskitalo, E. C. H. (2004) *Negotiating the Arctic. The Construction of an International Region*. Routledge, New York and London.

Keskitalo, E. C. H. and Liljenfeldt, J. (2014) Implementation of forest certification in Sweden: An issue of organisation and communication. *Scandinavian Journal of Forest Research* 29(5): 473–484.

Keskitalo, E. C. H., Pettersson, M. and Sörlin, S. (2019) Introduction. Understanding historical contingencies into the future: Cases from northern Europe. In: Keskitalo, E. C. H. (ed.) *The Politics of Arctic Resources: Change and Continuity in the 'Old North' of Northern Europe*. Routledge, London and New York. Pp. 1–17.

Keskitalo, E. C. H. and Preston, B. L. (2019) Understanding adaptation in the context of social theory. In: Keskitalo, E. C. H. and Preston, B. L. (eds.) *Research Handbook on Climate Change Adaptation Policy*. Edward Elgar, Cheltenham. Pp. 2–20.

Kingdon, J. W. (1995) *Agendas, Alternatives and Public Policies*. 2nd ed. HarperCollins, New York.

Kulynych, J. J. (1997) Performing politics: Foucault, Habermas, and postmodern participation. *Polity* 30(2): 315–346.

Lazlo Ambjörnsson, E., Keskitalo, E. C. H. and Karlsson, S. (2016) Forest discourses and the role of planning-related perspectives: The case of Sweden. *Scandinavian Journal of Forest Research* 31(1): 111–118.

Lisberg Jensen, E. (2002) *Som man ropar i skogen: Modernitet, makt och mångfald i kampen om Njakafjäll och i den svenska skogsbruksdebatten 1970–2000*. Lund University, Lund.

Mahoney, J. and Thelen, K. (2010) A theory of gradual institutional change. In: Mahoney, J. and Thelen, K. (eds.) *Explaining Institutional Change. Ambiguity, Agency, and Power*, Cambridge University Press, New York.

Manicas, P. T. (2006) *A Realist Philosophy of Social Science: Explanation and Understanding*. Cambridge University Press, Cambridge.

Mayes, C. R. (2015) Revisiting Foucault's 'normative confusions': Surveying the debate since the Collège de France lectures. *Philosophy Compass* 10(12): 841–855.

McNay, L. (1994) *Foucault. A Critical Introduction*. Polity Press, Cambridge.

Oels, A. (2019) The promise and limits of participation in adaptation governance: Moving beyond participation towards disruption. In: Keskitalo, E. C. H. and Preston, B. L. (eds.) *Research Handbook on Climate Change Adaptation Policy*. Edward Elgar, Cheltenham. Pp. 138–156.

Poster, M. (1984) *Foucault, Marxism and History. Mode of Production Versus Mode of Information*. Polity Press, Cambridge.

Ravetz, J. (2003) Models as metaphors. In: Kasemir, B., Jäger, J., Jaeger, C. C. and Gardner, M. T. (eds.) *Public Participation in Sustainability Science*. Cambridge University Press, Cambridge.

Rip, A. (1986) Controversies as informal technology assessment. *Knowledge: Creation, Diffusion, Utilization* 8(2): 349–371.

Rouse, J. (1996) Beyond epistemic sovereignty. In: Galison, P. and Stump, D. J. (eds.) *The Disunity of Science. Boundaries, Contexts, and Power*. Stanford University Press, Stanford.

Schön, D. A. and Rein, M. (1994) *Frame Reflection. Toward the Resolution of Intractable Policy Controversies*. Basic Books, New York.

Slovic, P. (1997) Trust, emotion, sex, politics, and science. Surveying the risk-assessment battlefield. In: Bazerman, M. H. et al. (eds.) *Environment, Ethics and Behavior: The Psychology of Environmental Valuation and Degradation*. The New Lexington Press, San Francisco.

Smart, B. (1983) *Foucault, Marxism and Critique*. Routledge and Kegan Paul, London and Boston.

Tenbrunsel, A. E. et al. (1997) Introduction. In: Bazerman, M. H. et al. (eds.) *Environment, Ethics and Behavior: The Psychology of Environmental Valuation and Degradation*. The New Lexington Press, San Francisco.

Thornton, P. H., Ocasio, W. and Lounsbury, M. (2012) *The Institutional Logics Perspective. A New Approach to Culture, Structure and Process*. Oxford University Press, Oxford.

Webler, T. (1995) "Right" discourse in citizen participation: An evaluative yardstick. In: Renn, O., Webler, T. and Wiedermann, P. (eds.) *Fairness and Competence in Citizen Participation. Evaluating Models for Environmental Discourse*. Kluwer Academic Publishers, Dordrecht and London.

Wuthnow, R., Hunter, J., Bergesen, A. and Kurzweil, E. (1984) *Cultural Analysis: The Work of Peter Berger, Mary Douglas, Michel Foucault, and Jürgen Habermas*. Routledge Kegan Paul, Boston.

Young, O. R. (with contributions from Agrawal, A., King, L. A., Sand, P. H., Underdal, A. and Wasson, M., revised version prepared by H. Schroeder). (1999) *Institutional Dimensions of Global Environmental Change. Science Plan*. IHDP, Bonn.

Chapter 6

Agrawal, A. (2010) Local institutions and adaptation to climate change. In: Mearns, R. and Norton, A. (eds) *Social Dimensions of Climate Change. Equity and Vulnerability in a Warming World*. World Bank, Washington, DC. Pp. 173–198.

Bachrach, P. and Botwinick, A. (1992) *Power and Empowerment. A Radical Theory of Participatory Democracy*. Temple University Press, Philadelphia.

Barnett, M. and Duvall, R. (2005) Power in international politics. *International Organization* 59(1): 39–75.

Beierle, T. C. (2002) The quality of stakeholder-based decisions. *Risk Analysis* 22(4): 739ff.

Benson, D., Lorenzoni, I. and Cook, H. (2016) Evaluating social learning in England flood risk management: An 'individual-community interaction' perspective. *Environmental Science & Policy* 55: 326–334.

Blake, J. (1999) Overcoming the 'value-action gap' in environmental policy: Tensions between national policy and local experience. *Local Environment* 4(3): 257ff.

Dahinden, U., Querol, C., Jäger, J. and Nilsson, M. (2003) Citizen interaction with computer models. In: Kasemir, B., Jäger, J., Jaeger, C. C. and Gardner, M. T. (eds.) *Public Participation in Sustainability Science*. Cambridge University Press, Cambridge.

Enserink, B. and Monnikhof, R. A. H. (2003) Information management for public participation in co-design processes: Evaluation of a Dutch example. *Journal of Environmental Planning and Management* 46(3): 315–344.

Eriksen, S. H., Nightingale, A. J. and Eakin, H. (2015) Reframing adaptation: The political nature of climate change adaptation. *Global Environmental Change* 35: 523–533.

Gough, C., et al. (2003) Contexts of citizen participation. In: Kasemir, B., Jäger, J., Jaeger, C. C. and Gardner, M. T. (eds.) *Public Participation in Sustainability Science*. Cambridge University Press, Cambridge.

Grothmann, T. (2011) Governance recommendations for adaptation in European urban regions: Results from five case studies and a European expert survey. In: Otto-Zimmermann, K. (ed.) *Resilient Cities: Cities and Adaptation to Climate Change*. Springer, Dordrecht. Pp. 167–176.

Gupta, J. (2008) Analysing scale and scaling in environmental governance. In: Young, O. R., King, L. A. and Schroeder, H. (eds.) *Institutions and Environmental Change: Principal Findings, Applications, and Research Frontiers*. Cambridge, MIT Press.

Hisschemöller, M., Tol, R. S. J. and Vellinga, P. (2001) The relevance of participatory approaches in integrated environmental assessment. *Integrated Assessment* 2: 57–72.

Holmes, T. and Scoones, I. (2000) Participatory Environmental Policy Processes: Experiences from North and South. IDS Working Paper 113. Institute of Development Studies, Brighton.

Hooghe, L. and Marks, G. (2001a) Types of multi-level governance. *European Integration online Papers (EIoP)* 5(11): 1–24.

Hooghe, L. and Marks, G. (2001b) *Multi-level Governance and European Integration*. Rowman & Littlefield, Boulder, CO.

Irwin, A. and Michael, M. (2003) *Science, Social Theory & Public Knowledge*. Open University Press, McGraw-Hill Education, Maidenhead.

Ivey, J. L., Smithers, J., de Löe, R. C. and Kreutzwizer, R. D. (2004) Community capacity for adaptation to climate-induced water shortages: Linking institutional complexity and local actors. *Environmental Management* 33(1): 36–47.

Kasemir, B., Jäger, J., Jaeger, C. C. and Gardner, M. T. (2003a) Preface. In: Kasemir, B., Jäger, J., Jaeger, C. C. and Gardner, M. T. (eds.) *Public Participation in Sustainability Science*. Cambridge University Press, Cambridge.

Kasemir, B., Jäger, J., Jaeger, C. C. and Gardner, M. T. (eds.) (2003b) *Public Participation in Sustainability Science*. Cambridge University Press, Cambridge.

Keskitalo, E. C. H. (2004) A framework for multi-level stakeholder studies in response to global change. *Local Environment* 9(5): 425–435.

Keskitalo, E. C. H. (2008) *Climate Change and Globalization in the Arctic: An Integrated Approach to Vulnerability Assessment*. Earthscan Publications, London. 257p.

Keskitalo, E. C. H. (2015) Actors' perceptions of issues in the implementation of the first round of the water framework directive: Examples from the water management and forestry sectors in southern Sweden. *Water* 7: 2202–2213.

Keskitalo, E. C. H. and Pettersson, M. (2012) 'Implementing multi-level governance? The legal basis and implementation of the EU water framework directive for forestry in Sweden. *Environmental Policy and Governance* 22(2), March/April: 90–103. Re-printed as Chapter 35 in Bache, I. and Flinders, M. (2015) *Multi-Level Governance: Essential Readings*. Elgar Research Reviews in Social and Political Science. Edward Elgar Press, Cheltenham.

Kochskämper, E., Challies, E., Newig, J. and Jager, N. W. (2016) Participation for effective environmental governance? Evidence from water framework directive implementation in Germany, Spain and the United Kingdom. *Journal of Environmental Management* 181: 737–748.

Lorenzoni, I., Jordan, A., Hulme, M., Turner, R. K. and O'Riordan, T. (2000) A co-evolutionary approach to climate change impact assessment: Part II. A scenario-based case study in East Anglia (UK). *Global Environmental Change* 10: 145–155.

Magnette, P. (2003) European governance and civic participation: Beyond elitist citizenship? *Political Studies* 51: 144–160.

Manicas, P. T. (2006) *A Realist Philosophy of Social Science: Explanation and Understanding*. Cambridge University Press, Cambridge.

Measham, T. G., Preston, B. L., Smith, T. F., Brooke, C., Gorddard, R., Withycombe, G. and Morrison, C. (2011) Adapting to climate change through local municipal planning: Barriers and challenges. *Mitigation and Adaptation Strategies for Global Change* 16: 889–909.

Moss, S., Pahl-Wostl, C. and Downing, T. (2001) Agent-based integrated assessment modelling: The example of climate change. *Integrated Assessment* 2: 17–30.

Moss, T. (2004) The governance of land use in river basins: Prospects for overcoming problems of institutional interplay with the EU water framework directive. *Land Use Policy* 21(1): 85–94.

Nanz, P. and Steffek, J. (2004) Global governance, participation and the public sphere. *Government and Opposition* 39(2): 314ff.

Owusu-Daaku, K. N. (2018) (Mal) Adaptation opportunism: When other interests take over stated or intended climate change adaptation objectives (and their unintended effects). *Local Environment* 23(9): 934–951.

Pateman, C. (1970) *Participation and Democratic Theory*. Cambridge University Press, Cambridge.

Pettersson, M., Stjernström, O. and Keskitalo, E. C. H. (2017) The role of participation in the planning process: Examples from Sweden. *Local Environment* 22(8): 986–997.

Philips, A. (1995) *The Politics of Presence*. Claredon Press, Oxford.

Plummer, R., Crona, B., Armitage, D., Olsson, P., Tengö, M. and Yudina, O. (2012) Adaptive comanagement: A systematic review and analysis. *Ecology and Society* 17(3).

Ravetz, J. (2003) Models as metaphors. In: Kasemir, B., Jäger, J., Jaeger, C. C. and Gardner, M. T. (eds.) *Public Participation in Sustainability Science*. Cambridge University Press, Cambridge.

Reed, J., Van Vianen, J., Deakin, E. L., Barlow, J. and Sunderland, T. (2016) Integrated landscape approaches to managing social and environmental issues in the tropics: Learning from the past to guide the future. *Global Change Biology* 22: 2540–2554.

Reed, M., Evely, A. C., Cundill, G., Fazey, I. R. A., Glass, J., Laing, A., Newig, J., Parrish, B., Prell, C., Raymond, C. and Stringer, L. (2010) What is social learning? *Ecology and Society* 15(4).

Schneider, S. H. (1997) Integrated assessment modelling of global climate change. Transparent rational tool for policy making or opaque screen hiding value-laden assumptions? *Environmental Modeling and Assessment* 2: 229–249.

Sisk, T. D., et al. (2001) *Democracy at the Local Level. The International IDEA Handbook on Participation, Representation, Conflict Management, and Governance.* International Institute for Democracy and Electoral Assistance IDEA, Stockholm.

Smith, J. B., Vogel, J. M. and Cromwell III, J. E. (2009) An architecture for government action on adaptation to climate change. An editorial comment. *Climatic Change* 95: 53–61.

Spyke, N. P. (1999) Public participation in environmental decision making at the new millennium: Structuring new spheres of public influence. *Boston College Environmental Affairs Law Review* 26(2): 263–314.

Stringer, L. C., Dougill, A. J., Fraser, E., Hubacek, K., Prell, C. and Reed, M. S. (2006) Unpacking "participation" in the adaptive management of social – ecological systems: A critical review. *Ecology and Society* 11(2): 39.

Tansey, J., Carmichael, J., Van Wynsberghe, R. and Robinson, J. (2002) The future is not what it used to be: Participatory integrated assessment in the Georgia Basin. *Global Environmental Change* 12(2): 97–104.

Urwin, K. and Jordan, A. (2008) Does public policy support or undermine climate change adaptation? Exploring policy interplay across different scales of governance. *Global Environmental Change* 18: 180–191.

Van Asselt, M. B. A. and Rijkens-Klomp, N. (2002) A look in the mirror: Reflection on participation in integrated assessment from a methodological perspective. *Global Environmental Change* 12: 167–184.

Van Asselt, M. B. A. and Rotmans, J. (2003) From projects to program in integrated assessment research. In: Kasemir, B., Jäger, J., Jaeger, C. C. and Gardner, M. T. (eds.) *Public Participation in Sustainability Science.* Cambridge University Press, Cambridge.

Van de Kerkhof, M. F. and Wieczorek, A. J. (2003) *On Transition Processes Towards Sustainability. A Methodological Perspective on the Involvement of Stakeholders.* (IVM Report; No. W-03/31). Department of Environmental Policy Analysis, Amsterdam.

Van den Bergh, J. C. (2013) Environmental and climate innovation: Limitations, policies and prices. *Technological Forecasting and Social Change* 80(1): 11–23.

Vulturius, G. and Gerger Swartling, Å. (2015) Overcoming social barriers to learning and engagement with climate change adaptation: Experiences with Swedish forestry stakeholders. *Scandinavian Journal of Forest Research* 30(3): 217–225.

Wesselink, A., Paavola, J., Fritsch, O. and Renn, O. (2011) Rationales for public participation in environmental policy and governance: Practitioners' perspectives. *Environment and Planning A* 43(11): 2688–2704.

Wynne, B. (1991) Knowledges in context. *Science, Technology & Human Values* 16(1): 111ff.

Chapter 7

ACIA (2005) *Arctic Climate Impact Assessment.* ACIA Overview report. Cambridge University Press, Cambridge. 1020p.

ACIA (2004) *Impacts of a Warming Arctic: Highlights.* Arctic Climate Impact Assessment. Cambridge University Press, Cambridge. 17p.

Andersson, E. and Keskitalo, E. C. H. (2018) Adaptation to climate change? Why business-as-usual remains the logical choice in Swedish forestry. *Global Environmental Change* 48(1): 76–85.

Back, A. and Marjavaara, R. (2017) Mapping an invisible population: The uneven geography of second-home tourism, *Tourism Geographies* 19(4): 595–611.

Beck, S. (2011) Moving beyond the linear model of expertise? IPCC and the test of adaptation. *Regional Environmental Change* 11(2): 297–306.

Bergstén, S. and Keskitalo, E. C. H. (2019) Feeling at home from a distance? How geographical distance and non-residency shape sense of place among private forest owners. *Society & Natural Resources* 32(2): 184–203.

Berkes, F., Colding, J. and Folke, C. (eds.) (2008) *Navigating Social-Ecological Systems: Building Resilience for Complexity and Change*. Cambridge University Press, Cambridge.

Borsboom, A. (1988) The savage in European social thought: A prelude to the conceptualization of the divergent peoples and cultures of Australia and Oceania. *Bijdragen tot de Taal-, Land-en Volkenkunde* 144(4): 419–432.

Brulle, R. J. and Dunlap, R. E. (2015) Sociology and climate change. In: Dunlap, R. E. and Brulle, R. J. (eds) *Climate Change and Society: Sociological perspectives*. Oxford University Press, Oxford. Pp. 1–21.

Büscher, B., van den Bremer, R., Fletcher, R. and Koot, S. (2017) Authenticity and the contradictions of the "ecotourism script": Global marketing and local politics in Ghana. *Critical Arts* 31(4): 37–52.

Carter, J. and Hollinsworth, D. (2009) Segregation and protectionism: Institutionalised views of Aboriginal rurality. *Journal of Rural Studies* 25: 414–424.

Cashore, B., Auld, G. and Newsom, D. (2004) *Governing Through Markets – Forest Certification and the Emergence of Nonstate Authority*. Yale University Press, New Haven.

Cashore, B. (2002) Legitimacy and the privatization of environmental governance: How non-state market-driven (NSMD) governance systems gain rule-making authority. *Governance* 15(4): 503–529.

Cote, M. and Nightingale, A. J. (2012) Resilience thinking meets social theory: Situating social change in socio-ecological systems (SES) research. *Progress in Human Geography* 36(4): 475–489.

Cronon, W. (1995) The myth of wilderness. In: Cronon, W. (ed.) *Uncommon Ground: Toward Reinventing Nature*. W.W. Norton, New York.

Cronon, W. (1996) The trouble with wilderness: Or getting back to the wrong nature. *Environmental History* 1(1): 7–28.

Cronon, W. (1987) Revisiting the vanishing frontier: The legacy of Frederick Jackson Turner. *The Western Historical Quarterly* 18(2): 157–176.

Cumming, G. S., Cumming, D. H. M. and Redman, C. L. (2006) Scale mismatches in social-ecological systems: Causes, consequences, and solutions. *Ecology and Society* 11: 14.

Descola, P. and Palsson, G. (eds). (1996) *Nature and Society: Anthropological Perspectives*. Routledge, London and New York.

Doel, R. E., Wråkberg, U. and Zeller, S. (2014) Science, environment, and the New arctic. *Journal of Historical Geography* 44: 2–14.

Dybbroe, S., Dahl, J. and Müller-Wille, L. (2010) Dynamics of Arctic urbanization. *Acta Borealia* 27(2): 120–124.

Elbakidze, M., Angelstam, P. K., Sandström, C. and Axelsson, R. (2010) Multi-stakeholder collaboration in Russian and Swedish model forest initiatives: Adaptive governance toward sustainable forest management? *Ecology and Society* 15(2): 14ff.

Erlandsson, E. (2016) The triad perspective on business models for wood harvesting. Diss. (sammanfattning/summary). Sveriges lantbruksuniv., Umeå. Acta Universitatis Agriculturae Sueciae, 1652–6880; 2016:124 [Doctoral thesis].

Ernstson, H., Barthel, S., Andersson, E. and Borgström, S. T. (2010) Scale-crossing brokers and network governance of urban ecosystem services: The case of Stockholm. *Ecology and Society* 15: 28.

Ford, J. D., Stephenson, E., Cunsolo Willox, A., Edge, V., Farahbakhsh, K., Furgal, C., . . . Austin, S. (2016) Community-based adaptation research in the Canadian Arctic. *Wiley Interdisciplinary Reviews: Climate Change* 7(2): 175–191.

Glacken, C. J. (1976) *Traces on the Rhodian Shore. Nature and culture in Western thought from ancient times to the end of the eighteenth century*. University of California Press, Berkeley.

Glackin, S. (2015) Contemporary urban culture: How community structures endure in an individualised society, *Culture and Organization* 21(1): 23–41.

Goodwin, I. (2012) Theorizing community as discourse in community informatics: "Resistant Identities" and contested technologies. *Communication and Critical/Cultural Studies* 9(1): 47–66.

Guerrero, A., McAllister, R., Corcoran, J. and Wilson, K. (2013) Scale mismatches, conservation planning, and the value of social network analyses. *Conservation Biology* 27: 35–44.

Gunnarsdotter, Y. (2005) *Från arbetsgemenskap till fritidsgemenskap. Den svenska landsbygdens omvandling ur Locknevibornas perspektiv Doctoral thesis No. 2005:3.* Faculty of Natural Resources and Agricultural Sciences. SLU, Uppsala.

Gupta, J. (2008) Analysing scale and scaling in environmental governance. In: Young, O. R., King, L. A. and Schroeder, H. (eds.) *Institutions and Environmental Change: Principal Findings, Applications, and Research Frontiers.* MIT Press, Cambridge.

Henderson, D. K. (1993) *Interpretation and Explanation in the Human Sciences.* SUNY Press, Albany.

Holling, C. S. and Gunderson, L. H. (2002) *Panarchy: Understanding Transformations in Human and Natural Systems.* Island Press, Washington, DC.

Holmes, T. and Scoones, I. (2000) *Participatory Environmental Policy Processes: Experiences from North and South.* IDS Working Paper 113. Institute of Development Studies, Brighton.

Hooghe, L. and Marks, G. (2001) Types of multi-level governance. *European Integration online Papers (EIoP)* 5(11): 1–24.

Hovelsrud, G. K. and Smit, B. (eds.) (2010) *Community Adaptation and Vulnerability in Arctic Regions.* Springer, Dordrecht.

Jørgensen, D. (2015) Rethinking rewilding. *Geoforum* 65: 482–488.

Jentoft, S. (2006) Beyond fisheries management: The Phronetic dimension. *Marine Policy* 30: 671–680.

Kagervall, A. (2014) On the conditions for developing hunting and fishing tourism in Sweden. Diss. (sammanfattning/summary). Sveriges lantbruksuniv., Umeå. Acta Universitatis Agriculturae Sueciae, 1652–6880; 2014:34 [Doctoral thesis].

Kashima, Y., Shi, J., Tsuchiya, K., Kashima, E. S., Cheng, S. Y. Y., Chao, M. M. and Shin, S.-H. (2011) Globalization and folk theory of social change: How globalization relates to societal perceptions about the past and future. *Journal of Social Issues* 67: 696–715.

Keskitalo, E. C. H., Lidestav, G., Westin, K. and Lindgren, U. (2020) Understanding the multiple dynamics of the countryside – Examples from forest cases in northern Europe. *Journal of Rural Studies* 78: 59–64.

Keskitalo, E. C. H., Lidestav, G., Karppinen, H. and Zivojinovic, I. (2017a) Is there a new European forest owner? The institutional context. In: Keskitalo, E. C. H. (ed.) *Globalisation and Change in Forest Ownership and Forest Use: Natural Resource Management in Transition.* Palgrave Macmillan, Basingstoke. Pp. 17–56.

Keskitalo, E. C. H., Karlsson, S., Lindgren, U., Pettersson, Ö., Lundmark, L., Slee, B., Villa, M. and Feliciano, D. (2017b) Rural – urban politics: Changing conceptions of the human-environment relationship. In: Keskitalo, E. C. H. (ed.) *Globalisation and Change in Forest Ownership and Forest Use: Natural Resource Management in Transition.* Palgrave Macmillan, Basingstoke. Pp. 183–224.

Keskitalo, E. C. H., Horstkotte, T., Kivinen, S., Forbes, B. and Käyhkö, J. (2016) "Generality of mis-fit?" The real-life difficulty of matching scales in an interconnected world. *Ambio* 45(6): 742–752.

Keskitalo, E. C. H. and Nuttall, M. (2015) Globalization of the 'Arctic'. In: Evengård, B., Nymand Larsen, J. (eds.) *The New Arctic*. Springer, Dordrecht. Pp. 175–187.

Keskitalo, E. C. H. (2008) *Climate Change and Globalization in the Arctic: An Integrated Approach to Vulnerability Assessment*. Earthscan Publications, London. 257p.

Keskitalo, E. C. H. (2004) *Negotiating the Arctic. The Construction of an International Region*. Routledge, New York and London. 282p.

Kingdon, J. W. (1995) *Agendas, Alternatives and Public Policies*. 2nd ed. HarperCollins, New York.

Kronholm, T. (2015) Forest owners' associations in a changing society. Diss. (sammanfattning/summary). Sveriges lantbruksuniv., Umeå. Acta Universitatis Agriculturae Sueciae, 1652–6880; 2015:102 [Doctoral thesis].

Kunnas, J., Keskitalo, E. C. H., Pettersson, M. and Stjernström, O. (2019) The institutionalisation of forestry as a primary land use in Sweden. In: Keskitalo, E. C. H. (ed.) *The Politics of Arctic Resources: Change and Continuity in the 'Old North' of Northern Europe*. Routledge, London and New York. Pp. 62–77.

Lafferty, W. M. and Cohen, F. (2001) Conclusions and perspectives. In: Lafferty, W. M. (ed.) *Sustainable Communities in Europe*. Earthscan, London.

Lidestav, G., Thellbro, C., Sandström, P., Lind, T., Holm, E., Olsson, O., Westin, K., Karppinen, H. and Ficko, A. (2017) Interactions between forest owners and their forests. In: Keskitalo, E. C. H. (ed.) *Globalisation and Change in Forest Ownership and Forest Use: Natural Resource Management in Transition*. Palgrave Macmillan, Basingstoke. Pp. 97–137.

Ljung, P. E., Widemo, F. and Ericsson, G. (2014) Trapping in predator management: Catching the profile of trap users in Sweden. *European Journal of Wildlife Research* 60: 681–689.

Macnaghten, P. and Urry, J. (1998) *Contested Natures*. Sage, London.

Marks, G. and Hooghe, L. (2004) Contrasting visions of multi-level governance. In: Bache, I. and Flinders, M. (eds.) *Multi-level Governance*. Oxford University Press, Oxford.

Mimura, N., Pulwarty, R. S., Duc, D. M., Elshinnawy, I., Redsteer, M. H., Huang, H. Q., Nkem, J. N. and Sanchez Rodriguez, R. A. (2014) Adaptation planning and implementation. In: Field, C. B., Barros, V. R., Dokken, D. J., Mach, K. J., Mastrandrea, M. D., Bilir, T. E., Chatterjee, M., Ebi, K. L., Estrada, Y. O., Genova, R. C., Girma, B., Kissel, E. S., Levy, A. N., MacCracken, S., Mastrandrea, P. R. and White, L. L. (eds.) *Climate Change 2014: Impacts, Adaptation, and Vulnerability. Part A: Global and Sectoral Aspects. Contribution of Working Group II to the Fifth Assessment Report of the Intergovernmental Panel on Climate Change*. Cambridge University Press, Cambridge, UK and New York. Pp. 869–898.

Nagendra, H. and Ostrom, E. (2012) Polycentric governance of multifunctional forested landscapes. *International Journal of the Commons* 6: 104–133.

Næss, L. O., Bang, G., Eriksen, S. and Vevatne, J. (2005) Institutional adaptation to climate change: Flood responses at the municipal level in Norway. *Global Environmental Change* 15: 125–138.

Nalau, J., Preston, B. L. and Maloney, M. C. (2015) Is adaptation a local responsibility? *Environmental Science & Policy* 48: 89–98.

Nanz, P. and Steffek, J. (2004) Global governance, participation and the public sphere. *Government and Opposition* 39(2): 314ff.

Nelson, M. P. and Callicott, J. B. (2008) *The Wilderness Debate Rages on: Continuing the Great New Wilderness Debate*. University of Georgia Press, London.

Noble, I. R., Huq, S., Anokhin, Y. A., Carmin, J., Goudou, D., Lansigan, F. P., Osman-Elasha, B. and Villamizar, A. (2014) Adaptation needs and options. In: Field, C. B., Barros, V. R., Dokken, D. J., Mach, K. J., Mastrandrea, M. D., Bilir, T. E., Chatterjee, M., Ebi,

K. L., Estrada, Y. O., Genova, R. C., Girma, B., Kissel, E. S., Levy, A. N., MacCracken, S., Mastrandrea, P. R. and White, L. L. (eds.) *Climate Change 2014: Impacts, Adaptation, and Vulnerability. Part A: Global and Sectoral Aspects. Contribution of Working Group II to the Fifth Assessment Report of the Intergovernmental Panel on Climate Change*. Cambridge University Press, Cambridge, UK and New York. Pp. 833–868.

Noble, I. (2019) The evolving interactions between adaptation research, international policy, and development practice. In: Keskitalo, E. C. H. and Preston, B. L. (eds.) *Research Handbook on Climate Change Adaptation Policy*. Edward Elgar, Cheltenham. Pp. 21–48.

Nordlund, A. and Westin, K. (2011) Forest values and forest management attitudes among private forest owners in Sweden. *Forests* 2(1): 30–50.

O'Brien, K. and Leichenko, R. (2000) Double exposure: Assessing the impacts of climate change within the context of economic globalization. *Global Environmental Change* 10: 221–232.

Olsson, L., Jerneck, A., Thoren, H., Persson, J. and O'Byrne, D. (2015) Why resilience is unappealing to social science: Theoretical and empirical investigations of the scientific use of resilience. *Science Advances* 1(4): e1400217.

Olsson, P., Folke, C. and Berkes, F. (2004) Adaptive comanagement for building resilience in social – ecological systems. *Environmental Management* 34(1): 75–90.

Palsson, G., Szerszynski, B., Sörlin, S., Marks, J., Avril, B., Crumley, C., . . . Buendía, M. P. (2013) Reconceptualizing the 'Anthropos' in the Anthropocene: Integrating the social sciences and humanities in global environmental change research. *Environmental Science & Policy* 28: 3–13.

Pateman, C. (1970) *Participation and Democratic Theory*. Cambridge University Press, Cambridge.

Pettersson, M., Stjernström, O. and Keskitalo, E. C. H. (2017) The role of participation in the planning process: Examples from Sweden. *Local Environment* 22(8): 986–997.

Philip, T. M., Way, W., Garcia, A. D., Schuler-Brown, S. and Navarro, O. (2013) When educators attempt to make "community" a part of classroom learning: The dangers of (mis)appropriating students' communities into schools. *Teaching and Teacher Education* 34: 174–183.

Pouta, E., Sievänen, T. and Neuvonen, M. (2006) Recreational wild berry picking in Finland – Reflection of a rural lifestyle. *Society & Natural Resources* 19: 285–304.

Prout, S. and Howitt, R. (2009) Frontier imaginings and subversive Indigenous spatialities. *Journal of Rural Studies* 25(4): 396–403.

Redclift, M. R. (2007) *Frontiers: Histories of Civil Society and Nature*. MIT Press, Cambridge.

Reed, M. G. (2019) The contributions of UNESCO Man and Biosphere Programme and biosphere reserves to the practice of sustainability science. *Sustainability Science* 14(3): 809–821.

Reed, J., Van Vianen, J., Deakin, E. L., Barlow, J. and Sunderland, T. (2016) Integrated landscape approaches to managing social and environmental issues in the tropics: Learning from the past to guide the future. *Global Change Biology* 22(7): 2540–2554.

Resilience Alliance. (2007) *Assessing Resilience in Social-Ecological Systems: A Workbook for Scientists*. Resilience Alliance, Stockholm.

Schraml, U. (2005) Between legitimacy and efficiency: The development of forestry associations in Germany. *Small-scale Forest Economics, Management and Policy* 4(3): 251–267.

Scott, M. (2008) Managing rural change and competing rationalities: Insights from conflicting rural storylines and local policy making in Ireland. *Planning Theory & Practice* 9(1): 9–32.

Sherval, M. (2009) Native Alaskan engagement with social constructions of rurality. *Journal of Rural Studies* 25: 425–434.

Silver, J. J. (2008) Weighing in on scale: Synthesizing disciplinary approaches to scale in the context of building interdisciplinary resource management. *Society & Natural Resources* 21(10): 921–929.

Sinclair, A. J. and Smith, D. L. (1999) Policy review: The model forest program in Canada: building consensus on sustainable forest management? *Society & Natural Resources* 12(2): 121–138.

Turner, F. J. (1921) *The Frontier in American History*. Henry Holt and Company, New York. [available through Project Gutenberg].

Tönnies, F. (1955) *Community and Association (Gemeinschaft und Gesellschaft)*. Routledge & Paul, London.

Von Bertalanffy, L. (1972) The history and status of general systems theory. *Academy of Management Journal* 15(4): 407–426.

von Essen, E., Hansen, H. P., Nordström Källström, H., Peterson, M. N. and Peterson, T. R. (2015) The radicalisation of rural resistance: How hunting counterpublics in the Nordic countries contribute to illegal hunting. *Journal of Rural Studies* 39: 199–209.

Ward, R. (1977) *The Australian legend*. Oxford University Press, Melbourne.

Westin, K., Eriksson, L., Lidestav, G., Karppinen, H., Haugen, K. and Nordlund, A. (2017) Individual forest owners in context. In: Keskitalo, E. C. H. (ed.) *Globalisation and Change in Forest Ownership and Forest Use: Natural Resource Management in Transition*. Palgrave Macmillan, Basingstoke. Pp. 57–95.

White, R. (1991) *'It's Your Misfortune and None of My Own'. A New History of the American West*. University of Oklahoma Press, Norman and London.

Chapter 8

Baker, S. and Eckerberg, K. (2007) Governance for sustainable development in Sweden: The experience of the local investment programme. *Local Environment* 12(4): 325–342.

Beck, S. (2011) Moving beyond the linear model of expertise? IPCC and the test of adaptation. *Regional Environmental Change* 11(2): 297–306.

Biagioli, M. (1996) From relativism to contingentism. In: Galison, P. and Stump, D. J. (eds.) *The Disunity of Science. Boundaries, Contexts, and Power*. Stanford University Press, Stanford.

Da Silva, S. E. and Shear, H. (2010) Great Lakes environmental indicators and state of the environment reporting: Use, needs, and limitations. *Local Environment* 15(8): 699–716.

Dunlap, R. E. and Brulle, R. J. (eds.) (2015) *Climate Change and Society: Sociological Perspectives*. Oxford University Press, Oxford.

Foucault, M. (1974) *The Archaeology of Knowledge*. Tavistock, London.

Funtowich, S. O. and Ravetz, J. R. (1990) *Uncertainty and Quality in Science for Policy*. Kluwer, Dordrecht.

Gupta, J. (2002) Global sustainable development governance: Institutional challenges from a theoretical perspective. *International Environmental Agreements* 2(4): 361–388.

Hollis, M. and Smith, S. (1991) *Explaining and Understanding International Relations*. Oxford University Press, Oxford.

Keskitalo, E. C. H. (2019) Conclusion. The "old North" – or quite simply the developed northern Europe. In: Keskitalo, E. C. H. (ed.) *The Politics of Arctic Resources: Change and Continuity in the 'Old North' of Northern Europe*. Routledge, London and New York. Pp. 242–262.

Keskitalo, E. C. H. and Andersson, E. (2017) Why organization may be the primary limitation to implementing sustainability at local level. Examples from Swedish case studies. *Resources* 6(1): 13ff.

Keskitalo, E. C. H. and Liljenfeldt, J. (2012) Working with sustainability. Experiences of sustainability processes in Swedish municipalities. *Natural Resources Forum* 36: 16–27.
Kingdon, J. W. (1995) *Agendas, Alternatives and Public Policies*. 2nd ed. HarperCollins, New York.
Konrad, K., Truffer, B. and Voss, J.-P. (2004) Transformation dynamics in utility systems an integrated approach to the analysis of transformation processes drawing on transition theory. In: Jacob, K., Binder, M. and Wieczorek, A. (eds.) *Governance for Industrial Transformation*. Proceedings of the 2003 Berlin Conference on the Human Dimensions of Global Environmental Change, Environmental Policy Research Centre, Berlin. Pp. 146–162.
Kuhn, T. S. (1970) *The Structure of Scientific Revolutions*. 2nd ed. [first edition 1962]. University of Chicago Press, Chicago.
Lafferty, W. M. and Cohen, F. (2001) Conclusions and perspectives. In: Lafferty, W. M. (ed.) *Sustainable Communities in Europe*. Earthscan, London.
Levy, M. A. (2003) The sustainable development research program. The effect of data and measurement limitations and strategies for overcoming them. Prepared for presentation at Open Meeting of Human Dimensions of Global Environmental Change Research Community Montreal, Canada 16–18 October.
Overland, I. and Sovacool, B. K. (2020) The misallocation of climate research funding. *Energy Research & Social Science* 62: 101349.
Rouse, J. (1996) Beyond epistemic sovereignty. In: Galison, P. and Stump, D. J. (eds.) *The Disunity of Science. Boundaries, Contexts, and Power*. Stanford University Press, Stanford.
Scheirer, M. A. (2005) Is sustainability possible? A review and commentary on empirical studies of program sustainability. *American Journal of Evaluation* 26(3): 320–347.
Voss, J.-P. (2003) Shaping socio-ecological transformation: The case for innovating governance. Paper presented at the Open Science Meeting of the International Human Dimensions Programme of Global Environmental Change Research. Montreal, 18 October.
Wellstead, A. M., Howlett, M. and Rayner, J. (2013) The neglect of governance in forest sector vulnerability assessments: Structural-functionalism and "black box" problems in climate change adaptation planning. *Ecology and Society* 18(3): 23.
Worster, D. (1996) The two cultures revisited: Environmental history and the environmental sciences. *Environment and History* 2(1): 3–14.

INDEX

Page numbers in *italics* indicate a figure on the corresponding page.